MW00562886

FREE Study Skills Videos/DVD Offer

Dear Customer,

Thank you for your purchase from Mometrix! We consider it an honor and a privilege that you have purchased our product and we want to ensure your satisfaction.

As part of our ongoing effort to meet the needs of test takers, we have developed a set of Study Skills Videos that we would like to give you for FREE. These videos cover our *best practices* for getting ready for your exam, from how to use our study materials to how to best prepare for the day of the test.

All that we ask is that you email us with feedback that would describe your experience so far with our product. Good, bad, or indifferent, we want to know what you think!

To get your FREE Study Skills Videos, you can use the **QR code** below, or send us an **email** at studyvideos@mometrix.com with *FREE VIDEOS* in the subject line and the following information in the body of the email:

- The name of the product you purchased.
- Your product rating on a scale of 1-5, with 5 being the highest rating.
- Your feedback. It can be long, short, or anything in between. We just want to know your impressions and experience so far with our product. (Good feedback might include how our study material met your needs and ways we might be able to make it even better. You could highlight features that you found helpful or features that you think we should add.)

If you have any questions or concerns, please don't hesitate to contact me directly.

Thanks again!

Sincerely,

Jay Willis
Vice President
jay.willis@mometrix.com
1-800-673-8175

SCAN HERE

Mometrix
TEST PREPARATION

Mometrix
TEST PREPARATION

IBEW

Aptitude Test Prep 2024-2025

Electrician Apprentice Secrets Study Guide

5 Full-Length Practice Tests

150+ Online Video Tutorials

Includes Detailed Answer Explanations

Copyright © 2024 by Mometrix Media LLC

All rights reserved. This product, or parts thereof, may not be reproduced, stored in a retrieval system, or transmitted in any form or by any means—electronic, mechanical, photocopy, recording, scanning, or other—except for brief quotations in critical reviews or articles, without the prior written permission of the publisher.

Written and edited by Matthew Bowling

Printed in the United States of America

This paper meets the requirements of ANSI/NISO Z39.48-1992 (Permanence of Paper).

Mometrix offers volume discount pricing to institutions. For more information or a price quote, please contact our sales department at sales@mometrix.com or 888-248-1219.

Mometrix Media LLC is not affiliated with or endorsed by any official testing organization. All organizational and test names are trademarks of their respective owners.

Paperback
ISBN 13: 978-1-5167-2642-4
ISBN 10: 1-5167-2642-1

DEAR FUTURE EXAM SUCCESS STORY

First of all, **THANK YOU** for purchasing Mometrix study materials!

Second, congratulations! You are one of the few determined test-takers who are committed to doing whatever it takes to excel on your exam. **You have come to the right place.** We developed these study materials with one goal in mind: to deliver you the information you need in a format that's concise and easy to use.

In addition to optimizing your guide for the content of the test, we've outlined our recommended steps for breaking down the preparation process into small, attainable goals so you can make sure you stay on track.

We've also analyzed the entire test-taking process, identifying the most common pitfalls and showing how you can overcome them and be ready for any curveball the test throws you.

Standardized testing is one of the biggest obstacles on your road to success, which only increases the importance of doing well in the high-pressure, high-stakes environment of test day. Your results on this test could have a significant impact on your future, and this guide provides the information and practical advice to help you achieve your full potential on test day.

Your success is our success

We would love to hear from you! If you would like to share the story of your exam success or if you have any questions or comments in regard to our products, please contact us at **800-673-8175** or **support@mometrix.com**.

Thanks again for your business and we wish you continued success!

Sincerely,
The Mometrix Test Preparation Team

Need more help? Check out our flashcards at:
http://mometrixflashcards.com/IBEW

Copyright © 2024 by Mometrix Media LLC. All rights reserved.
Written and edited by the Mometrix Exam Secrets Test Prep Team
Printed in the United States of America

Table of Contents

Introduction

Thank you for purchasing this resource! You have made the choice to prepare yourself for a test that could have a huge impact on your future, and this guide is designed to help you be fully ready for test day. Obviously, it's important to have a solid understanding of the test material, but you also need to be prepared for the unique environment and stressors of the test, so that you can perform to the best of your abilities.

For this purpose, the first section that appears in this guide is the **Secret Keys**. We've devoted countless hours to meticulously researching what works and what doesn't, and we've boiled down our findings to the five most impactful steps you can take to improve your performance on the test. We start at the beginning with study planning and move through the preparation process, all the way to the testing strategies that will help you get the most out of what you know when you're finally sitting in front of the test.

We recommend that you start preparing for your test as far in advance as possible. However, if you've bought this guide as a last-minute study resource and only have a few days before your test, we recommend that you skip over the first two Secret Keys since they address a long-term study plan.

If you struggle with **test anxiety**, we strongly encourage you to check out our recommendations for how you can overcome it. Test anxiety is a formidable foe, but it can be beaten, and we want to make sure you have the tools you need to defeat it.

Copyright © Mometrix Media. You have been licensed one copy of this document for personal use only. Any other reproduction or redistribution is strictly prohibited. All rights reserved.
This content is provided for test preparation purposes only and does not imply an endorsement by Mometrix of any particular political, scientific, or religious point of view.

Secret Key #1 – Plan Big, Study Small

There's a lot riding on your performance. If you want to ace this test, you're going to need to keep your skills sharp and the material fresh in your mind. You need a plan that lets you review everything you need to know while still fitting in your schedule. We'll break this strategy down into three categories.

Information Organization

Start with the information you already have: the official test outline. From this, you can make a complete list of all the concepts you need to cover before the test. Organize these concepts into groups that can be studied together, and create a list of any related vocabulary you need to learn so you can brush up on any difficult terms. You'll want to keep this vocabulary list handy once you actually start studying since you may need to add to it along the way.

Time Management

Once you have your set of study concepts, decide how to spread them out over the time you have left before the test. Break your study plan into small, clear goals so you have a manageable task for each day and know exactly what you're doing. Then just focus on one small step at a time. When you manage your time this way, you don't need to spend hours at a time studying. Studying a small block of content for a short period each day helps you retain information better and avoid stressing over how much you have left to do. You can relax knowing that you have a plan to cover everything in time. In order for this strategy to be effective though, you have to start studying early and stick to your schedule. Avoid the exhaustion and futility that comes from last-minute cramming!

Study Environment

The environment you study in has a big impact on your learning. Studying in a coffee shop, while probably more enjoyable, is not likely to be as fruitful as studying in a quiet room. It's important to keep distractions to a minimum. You're only planning to study for a short block of time, so make the most of it. Don't pause to check your phone or get up to find a snack. It's also important to **avoid multitasking**. Research has consistently shown that multitasking will make your studying dramatically less effective. Your study area should also be comfortable and well-lit so you don't have the distraction of straining your eyes or sitting on an uncomfortable chair.

The time of day you study is also important. You want to be rested and alert. Don't wait until just before bedtime. Study when you'll be most likely to comprehend and remember. Even better, if you know what time of day your test will be, set that time aside for study. That way your brain will be used to working on that subject at that specific time and you'll have a better chance of recalling information.

Finally, it can be helpful to team up with others who are studying for the same test. Your actual studying should be done in as isolated an environment as possible, but the work of organizing the information and setting up the study plan can be divided up. In between study sessions, you can discuss with your teammates the concepts that you're all studying and quiz each other on the details. Just be sure that your teammates are as serious about the test as you are. If you find that your study time is being replaced with social time, you might need to find a new team.

Copyright © Mometrix Media. You have been licensed one copy of this document for personal use only. Any other reproduction or redistribution is strictly prohibited. All rights reserved.
This content is provided for test preparation purposes only and does not imply an endorsement by Mometrix of any particular political, scientific, or religious point of view.

Secret Key #2 – Make Your Studying Count

You're devoting a lot of time and effort to preparing for this test, so you want to be absolutely certain it will pay off. This means doing more than just reading the content and hoping you can remember it on test day. It's important to make every minute of study count. There are two main areas you can focus on to make your studying count.

Retention

It doesn't matter how much time you study if you can't remember the material. You need to make sure you are retaining the concepts. To check your retention of the information you're learning, try recalling it at later times with minimal prompting. Try carrying around flashcards and glance at one or two from time to time or ask a friend who's also studying for the test to quiz you.

To enhance your retention, look for ways to put the information into practice so that you can apply it rather than simply recalling it. If you're using the information in practical ways, it will be much easier to remember. Similarly, it helps to solidify a concept in your mind if you're not only reading it to yourself but also explaining it to someone else. Ask a friend to let you teach them about a concept you're a little shaky on (or speak aloud to an imaginary audience if necessary). As you try to summarize, define, give examples, and answer your friend's questions, you'll understand the concepts better and they will stay with you longer. Finally, step back for a big picture view and ask yourself how each piece of information fits with the whole subject. When you link the different concepts together and see them working together as a whole, it's easier to remember the individual components.

Finally, practice showing your work on any multi-step problems, even if you're just studying. Writing out each step you take to solve a problem will help solidify the process in your mind, and you'll be more likely to remember it during the test.

Modality

Modality simply refers to the means or method by which you study. Choosing a study modality that fits your own individual learning style is crucial. No two people learn best in exactly the same way, so it's important to know your strengths and use them to your advantage.

For example, if you learn best by visualization, focus on visualizing a concept in your mind and draw an image or a diagram. Try color-coding your notes, illustrating them, or creating symbols that will trigger your mind to recall a learned concept. If you learn best by hearing or discussing information, find a study partner who learns the same way or read aloud to yourself. Think about how to put the information in your own words. Imagine that you are giving a lecture on the topic and record yourself so you can listen to it later.

For any learning style, flashcards can be helpful. Organize the information so you can take advantage of spare moments to review. Underline key words or phrases. Use different colors for different categories. Mnemonic devices (such as creating a short list in which every item starts with the same letter) can also help with retention. Find what works best for you and use it to store the information in your mind most effectively and easily.

Copyright © Mometrix Media. You have been licensed one copy of this document for personal use only. Any other reproduction or redistribution is strictly prohibited. All rights reserved.
This content is provided for test preparation purposes only and does not imply an endorsement by Mometrix of any particular political, scientific, or religious point of view.

Secret Key #3 – Practice the Right Way

Your success on test day depends not only on how many hours you put into preparing, but also on whether you prepared the right way. It's good to check along the way to see if your studying is paying off. One of the most effective ways to do this is by taking practice tests to evaluate your progress. Practice tests are useful because they show exactly where you need to improve. Every time you take a practice test, pay special attention to these three groups of questions:

- The questions you got wrong
- The questions you had to guess on, even if you guessed right
- The questions you found difficult or slow to work through

This will show you exactly what your weak areas are, and where you need to devote more study time. Ask yourself why each of these questions gave you trouble. Was it because you didn't understand the material? Was it because you didn't remember the vocabulary? Do you need more repetitions on this type of question to build speed and confidence? Dig into those questions and figure out how you can strengthen your weak areas as you go back to review the material.

Additionally, many practice tests have a section explaining the answer choices. It can be tempting to read the explanation and think that you now have a good understanding of the concept. However, an explanation likely only covers part of the question's broader context. Even if the explanation makes perfect sense, **go back and investigate** every concept related to the question until you're positive you have a thorough understanding.

As you go along, keep in mind that the practice test is just that: practice. Memorizing these questions and answers will not be very helpful on the actual test because it is unlikely to have any of the same exact questions. If you only know the right answers to the sample questions, you won't be prepared for the real thing. **Study the concepts** until you understand them fully, and then you'll be able to answer any question that shows up on the test.

It's important to wait on the practice tests until you're ready. If you take a test on your first day of study, you may be overwhelmed by the amount of material covered and how much you need to learn. Work up to it gradually.

On test day, you'll need to be prepared for answering questions, managing your time, and using the test-taking strategies you've learned. It's a lot to balance, like a mental marathon that will have a big impact on your future. Like training for a marathon, you'll need to start slowly and work your way up. When test day arrives, you'll be ready.

Start with the strategies you've read in the first two Secret Keys—plan your course and study in the way that works best for you. If you have time, consider using multiple study resources to get different approaches to the same concepts. It can be helpful to see difficult concepts from more than one angle. Then find a good source for practice tests. Many times, the test website will suggest potential study resources or provide sample tests.

Copyright © Mometrix Media. You have been licensed one copy of this document for personal use only. Any other reproduction or redistribution is strictly prohibited. All rights reserved.
This content is provided for test preparation purposes only and does not imply an endorsement by Mometrix of any particular political, scientific, or religious point of view.

Practice Test Strategy

If you're able to find at least three practice tests, we recommend this strategy:

UNTIMED AND OPEN-BOOK PRACTICE

Take the first test with no time constraints and with your notes and study guide handy. Take your time and focus on applying the strategies you've learned.

TIMED AND OPEN-BOOK PRACTICE

Take the second practice test open-book as well, but set a timer and practice pacing yourself to finish in time.

TIMED AND CLOSED-BOOK PRACTICE

Take any other practice tests as if it were test day. Set a timer and put away your study materials. Sit at a table or desk in a quiet room, imagine yourself at the testing center, and answer questions as quickly and accurately as possible.

Keep repeating timed and closed-book tests on a regular basis until you run out of practice tests or it's time for the actual test. Your mind will be ready for the schedule and stress of test day, and you'll be able to focus on recalling the material you've learned.

Copyright © Mometrix Media. You have been licensed one copy of this document for personal use only. Any other reproduction or redistribution is strictly prohibited. All rights reserved.
This content is provided for test preparation purposes only and does not imply an endorsement by Mometrix of any particular political, scientific, or religious point of view.

Secret Key #4 – Pace Yourself

Once you're fully prepared for the material on the test, your biggest challenge on test day will be managing your time. Just knowing that the clock is ticking can make you panic even if you have plenty of time left. Work on pacing yourself so you can build confidence against the time constraints of the exam. Pacing is a difficult skill to master, especially in a high-pressure environment, so **practice is vital.**

Set time expectations for your pace based on how much time is available. For example, if a section has 60 questions and the time limit is 30 minutes, you know you have to average 30 seconds or less per question in order to answer them all. Although 30 seconds is the hard limit, set 25 seconds per question as your goal, so you reserve extra time to spend on harder questions. When you budget extra time for the harder questions, you no longer have any reason to stress when those questions take longer to answer.

Don't let this time expectation distract you from working through the test at a calm, steady pace, but keep it in mind so you don't spend too much time on any one question. Recognize that taking extra time on one question you don't understand may keep you from answering two that you do understand later in the test. If your time limit for a question is up and you're still not sure of the answer, mark it and move on, and come back to it later if the time and the test format allow. If the testing format doesn't allow you to return to earlier questions, just make an educated guess; then put it out of your mind and move on.

On the easier questions, be careful not to rush. It may seem wise to hurry through them so you have more time for the challenging ones, but it's not worth missing one if you know the concept and just didn't take the time to read the question fully. Work efficiently but make sure you understand the question and have looked at all of the answer choices, since more than one may seem right at first.

Even if you're paying attention to the time, you may find yourself a little behind at some point. You should speed up to get back on track, but do so wisely. Don't panic; just take a few seconds less on each question until you're caught up. Don't guess without thinking, but do look through the answer choices and eliminate any you know are wrong. If you can get down to two choices, it is often worthwhile to guess from those. Once you've chosen an answer, move on and don't dwell on any that you skipped or had to hurry through. If a question was taking too long, chances are it was one of the harder ones, so you weren't as likely to get it right anyway.

On the other hand, if you find yourself getting ahead of schedule, it may be beneficial to slow down a little. The more quickly you work, the more likely you are to make a careless mistake that will affect your score. You've budgeted time for each question, so don't be afraid to spend that time. Practice an efficient but careful pace to get the most out of the time you have.

Copyright © Mometrix Media. You have been licensed one copy of this document for personal use only. Any other reproduction or redistribution is strictly prohibited. All rights reserved.
This content is provided for test preparation purposes only and does not imply an endorsement by Mometrix of any particular political, scientific, or religious point of view.

Secret Key #5 – Have a Plan for Guessing

When you're taking the test, you may find yourself stuck on a question. Some of the answer choices seem better than others, but you don't see the one answer choice that is obviously correct. What do you do?

The scenario described above is very common, yet most test takers have not effectively prepared for it. Developing and practicing a plan for guessing may be one of the single most effective uses of your time as you get ready for the exam.

In developing your plan for guessing, there are three questions to address:

- When should you start the guessing process?
- How should you narrow down the choices?
- Which answer should you choose?

When to Start the Guessing Process

Unless your plan for guessing is to select C every time (which, despite its merits, is not what we recommend), you need to leave yourself enough time to apply your answer elimination strategies. Since you have a limited amount of time for each question, that means that if you're going to give yourself the best shot at guessing correctly, you have to decide quickly whether or not you will guess.

Of course, the best-case scenario is that you don't have to guess at all, so first, see if you can answer the question based on your knowledge of the subject and basic reasoning skills. Focus on the key words in the question and try to jog your memory of related topics. Give yourself a chance to bring the knowledge to mind, but once you realize that you don't have (or you can't access) the knowledge you need to answer the question, it's time to start the guessing process.

It's almost always better to start the guessing process too early than too late. It only takes a few seconds to remember something and answer the question from knowledge. Carefully eliminating wrong answer choices takes longer. Plus, going through the process of eliminating answer choices can actually help jog your memory.

Summary: Start the guessing process as soon as you decide that you can't answer the question based on your knowledge.

Copyright © Mometrix Media. You have been licensed one copy of this document for personal use only. Any other reproduction or redistribution is strictly prohibited. All rights reserved.
This content is provided for test preparation purposes only and does not imply an endorsement by Mometrix of any particular political, scientific, or religious point of view.

How to Narrow Down the Choices

The next chapter in this book (**Test-Taking Strategies**) includes a wide range of strategies for how to approach questions and how to look for answer choices to eliminate. You will definitely want to read those carefully, practice them, and figure out which ones work best for you. Here though, we're going to address a mindset rather than a particular strategy.

Your odds of guessing an answer correctly depend on how many options you are choosing from.

Number of options left	5	4	3	2	1
Odds of guessing correctly	20%	25%	33%	50%	100%

You can see from this chart just how valuable it is to be able to eliminate incorrect answers and make an educated guess, but there are two things that many test takers do that cause them to miss out on the benefits of guessing:

- Accidentally eliminating the correct answer
- Selecting an answer based on an impression

We'll look at the first one here, and the second one in the next section.

To avoid accidentally eliminating the correct answer, we recommend a thought exercise called **the $5 challenge**. In this challenge, you only eliminate an answer choice from contention if you are willing to bet $5 on it being wrong. Why $5? Five dollars is a small but not insignificant amount of money. It's an amount you could afford to lose but wouldn't want to throw away. And while losing $5 once might not hurt too much, doing it twenty times will set you back $100. In the same way, each small decision you make—eliminating a choice here, guessing on a question there—won't by itself impact your score very much, but when you put them all together, they can make a big difference. By holding each answer choice elimination decision to a higher standard, you can reduce the risk of accidentally eliminating the correct answer.

The $5 challenge can also be applied in a positive sense: If you are willing to bet $5 that an answer choice *is* correct, go ahead and mark it as correct.

Summary: Only eliminate an answer choice if you are willing to bet $5 that it is wrong.

Copyright © Mometrix Media. You have been licensed one copy of this document for personal use only. Any other reproduction or redistribution is strictly prohibited. All rights reserved.
This content is provided for test preparation purposes only and does not imply an endorsement by Mometrix of any particular political, scientific, or religious point of view.

Which Answer to Choose

You're taking the test. You've run into a hard question and decided you'll have to guess. You've eliminated all the answer choices you're willing to bet $5 on. Now you have to pick an answer. Why do we even need to talk about this? Why can't you just pick whichever one you feel like when the time comes?

The answer to these questions is that if you don't come into the test with a plan, you'll rely on your impression to select an answer choice, and if you do that, you risk falling into a trap. The test writers know that everyone who takes their test will be guessing on some of the questions, so they intentionally write wrong answer choices to seem plausible. You still have to pick an answer though, and if the wrong answer choices are designed to look right, how can you ever be sure that you're not falling for their trap? The best solution we've found to this dilemma is to take the decision out of your hands entirely. Here is the process we recommend:

Once you've eliminated any choices that you are confident (willing to bet $5) are wrong, select the first remaining choice as your answer.

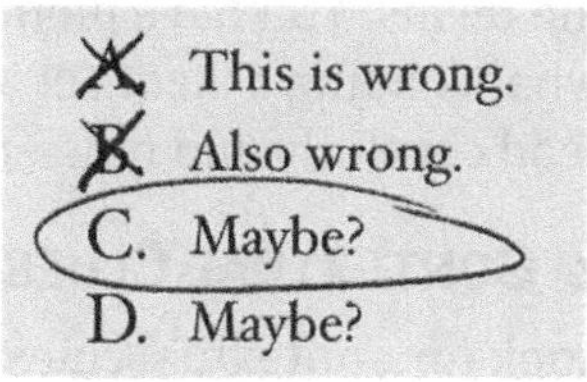

Whether you choose to select the first remaining choice, the second, or the last, the important thing is that you use some preselected standard. Using this approach guarantees that you will not be enticed into selecting an answer choice that looks right, because you are not basing your decision on how the answer choices look.

This is not meant to make you question your knowledge. Instead, it is to help you recognize the difference between your knowledge and your impressions. There's a huge difference between thinking an answer is right because of what you know, and thinking an answer is right because it looks or sounds like it should be right.

Summary: To ensure that your selection is appropriately random, make a predetermined selection from among all answer choices you have not eliminated.

Copyright © Mometrix Media. You have been licensed one copy of this document for personal use only. Any other reproduction or redistribution is strictly prohibited. All rights reserved.
This content is provided for test preparation purposes only and does not imply an endorsement by Mometrix of any particular political, scientific, or religious point of view.

Test-Taking Strategies

This section contains a list of test-taking strategies that you may find helpful as you work through the test. By taking what you know and applying logical thought, you can maximize your chances of answering any question correctly!

It is very important to realize that every question is different and every person is different: no single strategy will work on every question, and no single strategy will work for every person. That's why we've included all of them here, so you can try them out and determine which ones work best for different types of questions and which ones work best for you.

Question Strategies

⊘ Read Carefully

Read the question and the answer choices carefully. Don't miss the question because you misread the terms. You have plenty of time to read each question thoroughly and make sure you understand what is being asked. Yet a happy medium must be attained, so don't waste too much time. You must read carefully and efficiently.

⊘ Contextual Clues

Look for contextual clues. If the question includes a word you are not familiar with, look at the immediate context for some indication of what the word might mean. Contextual clues can often give you all the information you need to decipher the meaning of an unfamiliar word. Even if you can't determine the meaning, you may be able to narrow down the possibilities enough to make a solid guess at the answer to the question.

⊘ Prefixes

If you're having trouble with a word in the question or answer choices, try dissecting it. Take advantage of every clue that the word might include. Prefixes can be a huge help. Usually, they allow you to determine a basic meaning. *Pre-* means before, *post-* means after, *pro-* is positive, *de-* is negative. From prefixes, you can get an idea of the general meaning of the word and try to put it into context.

⊘ Hedge Words

Watch out for critical hedge words, such as *likely, may, can, sometimes, often, almost, mostly, usually, generally, rarely,* and *sometimes.* Question writers insert these hedge phrases to cover every possibility. Often an answer choice will be wrong simply because it leaves no room for exception. Be on guard for answer choices that have definitive words such as *exactly* and *always.*

⊘ Switchback Words

Stay alert for *switchbacks.* These are the words and phrases frequently used to alert you to shifts in thought. The most common switchback words are *but, although,* and *however.* Others include *nevertheless, on the other hand, even though, while, in spite of, despite,* and *regardless of.* Switchback words are important to catch because they can change the direction of the question or an answer choice.

Copyright © Mometrix Media. You have been licensed one copy of this document for personal use only. Any other reproduction or redistribution is strictly prohibited. All rights reserved.
This content is provided for test preparation purposes only and does not imply an endorsement by Mometrix of any particular political, scientific, or religious point of view.

✓ Face Value

When in doubt, use common sense. Accept the situation in the problem at face value. Don't read too much into it. These problems will not require you to make wild assumptions. If you have to go beyond creativity and warp time or space in order to have an answer choice fit the question, then you should move on and consider the other answer choices. These are normal problems rooted in reality. The applicable relationship or explanation may not be readily apparent, but it is there for you to figure out. Use your common sense to interpret anything that isn't clear.

Answer Choice Strategies

✓ Answer Selection

The most thorough way to pick an answer choice is to identify and eliminate wrong answers until only one is left, then confirm it is the correct answer. Sometimes an answer choice may immediately seem right, but be careful. The test writers will usually put more than one reasonable answer choice on each question, so take a second to read all of them and make sure that the other choices are not equally obvious. As long as you have time left, it is better to read every answer choice than to pick the first one that looks right without checking the others.

✓ Answer Choice Families

An answer choice family consists of two (in rare cases, three) answer choices that are very similar in construction and cannot all be true at the same time. If you see two answer choices that are direct opposites or parallels, one of them is usually the correct answer. For instance, if one answer choice says that quantity x increases and another either says that quantity x decreases (opposite) or says that quantity y increases (parallel), then those answer choices would fall into the same family. An answer choice that doesn't match the construction of the answer choice family is more likely to be incorrect. Most questions will not have answer choice families, but when they do appear, you should be prepared to recognize them.

✓ Eliminate Answers

Eliminate answer choices as soon as you realize they are wrong, but make sure you consider all possibilities. If you are eliminating answer choices and realize that the last one you are left with is also wrong, don't panic. Start over and consider each choice again. There may be something you missed the first time that you will realize on the second pass.

✓ Avoid Fact Traps

Don't be distracted by an answer choice that is factually true but doesn't answer the question. You are looking for the choice that answers the question. Stay focused on what the question is asking for so you don't accidentally pick an answer that is true but incorrect. Always go back to the question and make sure the answer choice you've selected actually answers the question and is not merely a true statement.

✓ Extreme Statements

In general, you should avoid answers that put forth extreme actions as standard practice or proclaim controversial ideas as established fact. An answer choice that states the "process should be used in certain situations, if..." is much more likely to be correct than one that states the "process should be discontinued completely." The first is a calm rational statement and doesn't even make a definitive, uncompromising stance, using a hedge word *if* to provide wiggle room, whereas the second choice is far more extreme.

Copyright © Mometrix Media. You have been licensed one copy of this document for personal use only. Any other reproduction or redistribution is strictly prohibited. All rights reserved.
This content is provided for test preparation purposes only and does not imply an endorsement by Mometrix of any particular political, scientific, or religious point of view.

✓ Benchmark

As you read through the answer choices and you come across one that seems to answer the question well, mentally select that answer choice. This is not your final answer, but it's the one that will help you evaluate the other answer choices. The one that you selected is your benchmark or standard for judging each of the other answer choices. Every other answer choice must be compared to your benchmark. That choice is correct until proven otherwise by another answer choice beating it. If you find a better answer, then that one becomes your new benchmark. Once you've decided that no other choice answers the question as well as your benchmark, you have your final answer.

✓ Predict the Answer

Before you even start looking at the answer choices, it is often best to try to predict the answer. When you come up with the answer on your own, it is easier to avoid distractions and traps because you will know exactly what to look for. The right answer choice is unlikely to be word-for-word what you came up with, but it should be a close match. Even if you are confident that you have the right answer, you should still take the time to read each option before moving on.

General Strategies

✓ Tough Questions

If you are stumped on a problem or it appears too hard or too difficult, don't waste time. Move on! Remember though, if you can quickly check for obviously incorrect answer choices, your chances of guessing correctly are greatly improved. Before you completely give up, at least try to knock out a couple of possible answers. Eliminate what you can and then guess at the remaining answer choices before moving on.

✓ Check Your Work

Since you will probably not know every term listed and the answer to every question, it is important that you get credit for the ones that you do know. Don't miss any questions through careless mistakes. If at all possible, try to take a second to look back over your answer selection and make sure you've selected the correct answer choice and haven't made a costly careless mistake (such as marking an answer choice that you didn't mean to mark). This quick double check should more than pay for itself in caught mistakes for the time it costs.

✓ Pace Yourself

It's easy to be overwhelmed when you're looking at a page full of questions; your mind is confused and full of random thoughts, and the clock is ticking down faster than you would like. Calm down and maintain the pace that you have set for yourself. Especially as you get down to the last few minutes of the test, don't let the small numbers on the clock make you panic. As long as you are on track by monitoring your pace, you are guaranteed to have time for each question.

✓ Don't Rush

It is very easy to make errors when you are in a hurry. Maintaining a fast pace in answering questions is pointless if it makes you miss questions that you would have gotten right otherwise. Test writers like to include distracting information and wrong answers that seem right. Taking a little extra time to avoid careless mistakes can make all the difference in your test score. Find a pace that allows you to be confident in the answers that you select.

Copyright © Mometrix Media. You have been licensed one copy of this document for personal use only. Any other reproduction or redistribution is strictly prohibited. All rights reserved.
This content is provided for test preparation purposes only and does not imply an endorsement by Mometrix of any particular political, scientific, or religious point of view.

⊘ Keep Moving

Panicking will not help you pass the test, so do your best to stay calm and keep moving. Taking deep breaths and going through the answer elimination steps you practiced can help to break through a stress barrier and keep your pace.

Final Notes

The combination of a solid foundation of content knowledge and the confidence that comes from practicing your plan for applying that knowledge is the key to maximizing your performance on test day. As your foundation of content knowledge is built up and strengthened, you'll find that the strategies included in this chapter become more and more effective in helping you quickly sift through the distractions and traps of the test to isolate the correct answer.

Now that you're preparing to move forward into the test content chapters of this book, be sure to keep your goal in mind. As you read, think about how you will be able to apply this information on the test. If you've already seen sample questions for the test and you have an idea of the question format and style, try to come up with questions of your own that you can answer based on what you're reading. This will give you valuable practice applying your knowledge in the same ways you can expect to on test day.

Good luck and good studying!

Copyright © Mometrix Media. You have been licensed one copy of this document for personal use only. Any other reproduction or redistribution is strictly prohibited. All rights reserved.
This content is provided for test preparation purposes only and does not imply an endorsement by Mometrix of any particular political, scientific, or religious point of view.

Copyright © Mometrix Media. You have been licensed one copy of this document for personal use only. Any other reproduction or redistribution is strictly prohibited. All rights reserved.
This content is provided for test preparation purposes only and does not imply an endorsement by Mometrix of any particular political, scientific, or religious point of view.

Algebra and Functions

Number Basics

Classifications of Numbers

Numbers are the basic building blocks of mathematics. Specific features of numbers are identified by the following terms:

Integer – any positive or negative whole number, including zero. Integers do not include fractions $\left(\frac{1}{3}\right)$, decimals (0.56), or mixed numbers $\left(7\frac{3}{4}\right)$.

Prime number – any whole number greater than 1 that has only two factors, itself and 1; that is, a number that can be divided evenly only by 1 and itself.

Composite number – any whole number greater than 1 that has more than two different factors; in other words, any whole number that is not a prime number. For example: The composite number 8 has the factors of 1, 2, 4, and 8.

Even number – any integer that can be divided by 2 without leaving a remainder. For example: 2, 4, 6, 8, and so on.

Odd number – any integer that cannot be divided evenly by 2. For example: 3, 5, 7, 9, and so on.

Decimal number – any number that uses a decimal point to show the part of the number that is less than one. Example: 1.234.

Decimal point – a symbol used to separate the ones place from the tenths place in decimals or dollars from cents in currency.

Decimal place – the position of a number to the right of the decimal point. In the decimal 0.123, the 1 is in the first place to the right of the decimal point, indicating tenths; the 2 is in the second place, indicating hundredths; and the 3 is in the third place, indicating thousandths.

The **decimal**, or base 10, system is a number system that uses ten different digits (0, 1, 2, 3, 4, 5, 6, 7, 8, 9). An example of a number system that uses something other than ten digits is the **binary**, or base 2, number system, used by computers, which uses only the numbers 0 and 1. It is thought that the decimal system originated because people had only their 10 fingers for counting.

Rational numbers include all integers, decimals, and fractions. Any terminating or repeating decimal number is a rational number.

Irrational numbers cannot be written as fractions or decimals because the number of decimal places is infinite and there is no recurring pattern of digits within the number. For example, pi (π) begins with 3.141592 and continues without terminating or repeating, so pi is an irrational number.

Copyright © Mometrix Media. You have been licensed one copy of this document for personal use only. Any other reproduction or redistribution is strictly prohibited. All rights reserved.
This content is provided for test preparation purposes only and does not imply an endorsement by Mometrix of any particular political, scientific, or religious point of view.

Real numbers are the set of all rational and irrational numbers.

> **Review Video: Classification of Numbers**
> Visit mometrix.com/academy and enter code: 461071
>
> **Review Video: Prime and Composite Numbers**
> Visit mometrix.com/academy and enter code: 565581

NUMBERS IN WORD FORM AND PLACE VALUE

When writing numbers out in word form or translating word form to numbers, it is essential to understand how a place value system works. In the decimal or base-10 system, each digit of a number represents how many of the corresponding place value—a specific factor of 10—are contained in the number being represented. To make reading numbers easier, every three digits to the left of the decimal place is preceded by a comma. The following table demonstrates some of the place values:

Power of 10	10^3	10^2	10^1	10^0	10^{-1}	10^{-2}	10^{-3}
Value	1,000	100	10	1	0.1	0.01	0.001
Place	thousands	hundreds	tens	ones	tenths	hundredths	thousandths

For example, consider the number 4,546.09, which can be separated into each place value like this:

4: thousands
5: hundreds
4: tens
6: ones
0: tenths
9: hundredths

This number in word form would be *four thousand five hundred forty-six and nine hundredths.*

> **Review Video: Place Value**
> Visit mometrix.com/academy and enter code: 205433

Rational Numbers

The term **rational** means that the number can be expressed as a ratio or fraction. That is, a number, r, is rational if and only if it can be represented by a fraction $\frac{a}{b}$ where a and b are integers and b does not equal 0. The set of rational numbers includes integers and decimals. If there is no finite way to represent a value with a fraction of integers, then the number is **irrational**. Common examples of irrational numbers include: $\sqrt{5}$, $(1 + \sqrt{2})$, and π.

> **Review Video: Rational and Irrational Numbers**
> Visit mometrix.com/academy and enter code: 280645
>
> **Review Video: Ordering Rational Numbers**
> Visit mometrix.com/academy and enter code: 419578

Copyright © Mometrix Media. You have been licensed one copy of this document for personal use only. Any other reproduction or redistribution is strictly prohibited. All rights reserved.
This content is provided for test preparation purposes only and does not imply an endorsement by Mometrix of any particular political, scientific, or religious point of view.

Number Lines

A number line is a graph to see the distance between numbers. Basically, this graph shows the relationship between numbers. So a number line may have a point for zero and may show negative numbers on the left side of the line. Any positive numbers are placed on the right side of the line. For example, consider the points labeled on the following number line:

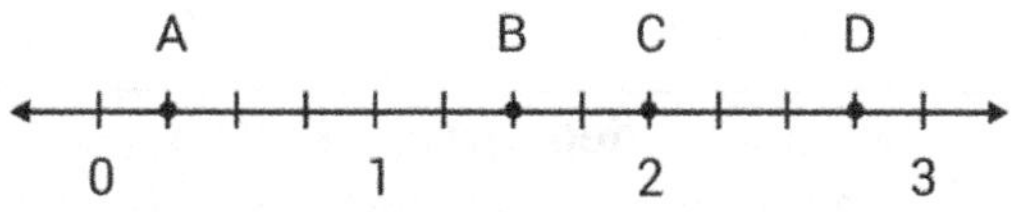

We can use the dashed lines on the number line to identify each point. Each dashed line between two whole numbers is $\frac{1}{4}$. The line halfway between two numbers is $\frac{1}{2}$.

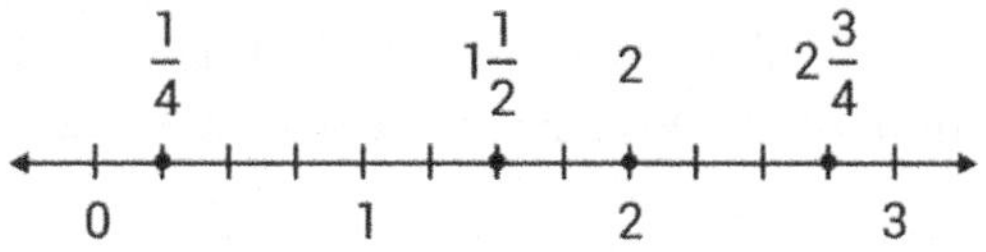

Review Video: The Number Line
Visit mometrix.com/academy and enter code: 816439

Rounding and Estimation

Rounding is reducing the digits in a number while still trying to keep the value similar. The result will be less accurate but in a simpler form and easier to use. Whole numbers can be rounded to the nearest ten, hundred, or thousand.

When you are asked to estimate the solution to a problem, you will need to provide only an approximate figure or **estimation** for your answer. In this situation, you will need to round each number in the calculation to the level indicated (nearest hundred, nearest thousand, etc.) or to a level that makes sense for the numbers involved. When estimating a sum **all numbers must be rounded to the same level**. You cannot round one number to the nearest thousand while rounding another to the nearest hundred.

Review Video: Rounding and Estimation
Visit mometrix.com/academy and enter code: 126243

Absolute Value

A precursor to working with negative numbers is understanding what **absolute values** are. A number's absolute value is simply the distance away from zero a number is on the number line. The absolute value of a number is always positive and is written $|x|$. For example, the absolute value of 3, written as $|3|$, is 3 because the distance between 0 and 3 on a number line is three units.

Copyright © Mometrix Media. You have been licensed one copy of this document for personal use only. Any other reproduction or redistribution is strictly prohibited. All rights reserved.
This content is provided for test preparation purposes only and does not imply an endorsement by Mometrix of any particular political, scientific, or religious point of view.

Likewise, the absolute value of -3, written as $|-3|$, is 3 because the distance between 0 and -3 on a number line is three units. So $|3| = |-3|$.

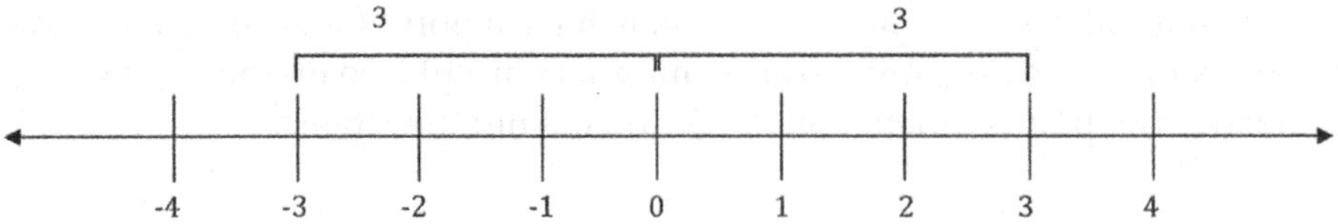

Review Video: Absolute Value
Visit mometrix.com/academy and enter code: 314669

Operations

An **operation** is simply a mathematical process that takes some value(s) as input(s) and produces an output. Elementary operations are often written in the following form: *value operation value*. For instance, in the expression $1 + 2$ the values are 1 and 2 and the operation is addition. Performing the operation gives the output of 3. In this way we can say that $1 + 2$ and 3 are equal, or $1 + 2 = 3$.

Addition

Addition increases the value of one quantity by the value of another quantity (both called **addends**). Example: $2 + 4 = 6$ or $8 + 9 = 17$. The result is called the **sum**. With addition, the order does not matter, $4 + 2 = 2 + 4$.

When adding signed numbers, if the signs are the same simply add the absolute values of the addends and apply the original sign to the sum. For example, $(+4) + (+8) = +12$ and $(-4) + (-8) = -12$. When the original signs are different, take the absolute values of the addends and subtract the smaller value from the larger value, then apply the original sign of the larger value to the difference. Example: $(+4) + (-8) = -4$ and $(-4) + (+8) = +4$.

Subtraction

Subtraction is the opposite operation to addition; it decreases the value of one quantity (the **minuend**) by the value of another quantity (the **subtrahend**). For example, $6 - 4 = 2$ or $17 - 8 = 9$. The result is called the **difference**. Note that with subtraction, the order does matter, $6 - 4 \neq 4 - 6$.

For subtracting signed numbers, change the sign of the subtrahend and then follow the same rules used for addition. Example: $(+4) - (+8) = (+4) + (-8) = -4$

Multiplication

Multiplication can be thought of as repeated addition. One number (the **multiplier**) indicates how many times to add the other number (the **multiplicand**) to itself. Example: $3 \times 2 = 2 + 2 + 2 = 6$. With multiplication, the order does not matter, $2 \times 3 = 3 \times 2$ or $3 + 3 = 2 + 2 + 2$, either way the result (the **product**) is the same.

If the signs are the same, the product is positive when multiplying signed numbers. Example: $(+4) \times (+8) = +32$ and $(-4) \times (-8) = +32$. If the signs are opposite, the product is negative. Example: $(+4) \times (-8) = -32$ and $(-4) \times (+8) = -32$. When more than two factors are multiplied together, the sign of the product is determined by how many negative factors are present. If there are an odd number of negative factors then the product is negative, whereas an even number of

Copyright © Mometrix Media. You have been licensed one copy of this document for personal use only. Any other reproduction or redistribution is strictly prohibited. All rights reserved.
This content is provided for test preparation purposes only and does not imply an endorsement by Mometrix of any particular political, scientific, or religious point of view.

negative factors indicates a positive product. Example: $(+4) \times (-8) \times (-2) = +64$ and $(-4) \times (-8) \times (-2) = -64$.

DIVISION

Division is the opposite operation to multiplication; one number (the **divisor**) tells us how many parts to divide the other number (the **dividend**) into. The result of division is called the **quotient**. Example: $20 \div 4 = 5$. If 20 is split into 4 equal parts, each part is 5. With division, the order of the numbers does matter, $20 \div 4 \neq 4 \div 20$.

The rules for dividing signed numbers are similar to multiplying signed numbers. If the dividend and divisor have the same sign, the quotient is positive. If the dividend and divisor have opposite signs, the quotient is negative. Example: $(-4) \div (+8) = -0.5$.

Review Video: Mathematical Operations
Visit mometrix.com/academy and enter code: 208095

PARENTHESES

Parentheses are used to designate which operations should be done first when there are multiple operations. Example: $4 - (2 + 1) = 1$; the parentheses tell us that we must add 2 and 1, and then subtract the sum from 4, rather than subtracting 2 from 4 and then adding 1 (this would give us an answer of 3).

Review Video: Mathematical Parentheses
Visit mometrix.com/academy and enter code: 978600

EXPONENTS

An **exponent** is a superscript number placed next to another number at the top right. It indicates how many times the base number is to be multiplied by itself. Exponents provide a shorthand way to write what would be a longer mathematical expression, Example: $2^4 = 2 \times 2 \times 2 \times 2$. A number with an exponent of 2 is said to be "squared," while a number with an exponent of 3 is said to be "cubed." The value of a number raised to an exponent is called its power. So 8^4 is read as "8 to the 4th power," or "8 raised to the power of 4."

Review Video: Exponents
Visit mometrix.com/academy and enter code: 600998

ROOTS

A **root**, such as a square root, is another way of writing a fractional exponent. Instead of using a superscript, roots use the radical symbol ($\sqrt{}$) to indicate the operation. A radical will have a number underneath the bar, and may sometimes have a number in the upper left: $\sqrt[n]{a}$, read as "the nth root of a." The relationship between radical notation and exponent notation can be described by this equation:

$$\sqrt[n]{a} = a^{\frac{1}{n}}$$

The two special cases of $n = 2$ and $n = 3$ are called square roots and cube roots. If there is no number to the upper left, the radical is understood to be a square root ($n = 2$). Nearly all of the roots you encounter will be square roots. A square root is the same as a number raised to the one-

Copyright © Mometrix Media. You have been licensed one copy of this document for personal use only. Any other reproduction or redistribution is strictly prohibited. All rights reserved.
This content is provided for test preparation purposes only and does not imply an endorsement by Mometrix of any particular political, scientific, or religious point of view.

half power. When we say that a is the square root of b ($a = \sqrt{b}$), we mean that a multiplied by itself equals b: ($a \times a = b$).

A **perfect square** is a number that has an integer for its square root. There are 10 perfect squares from 1 to 100: 1, 4, 9, 16, 25, 36, 49, 64, 81, 100 (the squares of integers 1 through 10).

Review Video: Roots
Visit mometrix.com/academy and enter code: 795655

Review Video: Perfect Squares and Square Roots
Visit mometrix.com/academy and enter code: 648063

Word Problems and Mathematical Symbols

When working on word problems, you must be able to translate verbal expressions or "math words" into math symbols. This chart contains several "math words" and their appropriate symbols:

Phrase	Symbol
equal, is, was, will be, has, costs, gets to, is the same as, becomes	$=$
times, of, multiplied by, product of, twice, doubles, halves, triples	$\times$
divided by, per, ratio of/to, out of	$\div$
plus, added to, sum, combined, and, more than, totals of	$+$
subtracted from, less than, decreased by, minus, difference between	$-$
what, how much, original value, how many, a number, a variable	x, n, etc.

Examples of Translated Mathematical Phrases

- The phrase four more than twice a number can be written algebraically as $2x + 4$.
- The phrase half a number decreased by six can be written algebraically as $\frac{1}{2}x - 6$.
- The phrase the sum of a number and the product of five and that number can be written algebraically as $x + 5x$.
- You may see a test question that says, "Olivia is constructing a bookcase from seven boards. Two of them are for vertical supports and five are for shelves. The height of the bookcase is twice the width of the bookcase. If the seven boards total 36 feet in length, what will be the height of Olivia's bookcase?" You would need to make a sketch and then create the equation to determine the width of the shelves. The height can be represented as double the width. (If x represents the width of the shelves in feet, then the height of the bookcase is $2x$. Since the seven boards total 36 feet, $2x + 2x + x + x + x + x + x = 36$ or $9x = 36$; $x = 4$. The height is twice the width, or 8 feet.)

Subtraction with Regrouping

A great way to make use of some of the features built into the decimal system would be regrouping when attempting longform subtraction operations. When subtracting within a place value, sometimes the minuend is smaller than the subtrahend, **regrouping** enables you to 'borrow' a unit

Copyright © Mometrix Media. You have been licensed one copy of this document for personal use only. Any other reproduction or redistribution is strictly prohibited. All rights reserved.
This content is provided for test preparation purposes only and does not imply an endorsement by Mometrix of any particular political, scientific, or religious point of view.

from a place value to the left in order to get a positive difference. For example, consider subtracting 189 from 525 with regrouping.

First, set up the subtraction problem in vertical form:

$$\begin{array}{r} 525 \\ -\ 189 \\ \hline \end{array}$$

Notice that the numbers in the ones and tens columns of 525 are smaller than the numbers in the ones and tens columns of 189. This means you will need to use regrouping to perform subtraction:

$$\begin{array}{cccc} & 5 & 2 & 5 \\ - & 1 & 8 & 9 \\ \hline \end{array}$$

To subtract 9 from 5 in the ones column you will need to borrow from the 2 in the tens columns:

$$\begin{array}{cccc} & 5 & 1 & 15 \\ - & 1 & 8 & 9 \\ \hline & & & 6 \end{array}$$

Next, to subtract 8 from 1 in the tens column you will need to borrow from the 5 in the hundreds column:

$$\begin{array}{cccc} & 4 & 11 & 15 \\ - & 1 & 8 & 9 \\ \hline & & 3 & 6 \end{array}$$

Last, subtract the 1 from the 4 in the hundreds column:

$$\begin{array}{cccc} & 4 & 11 & 15 \\ - & 1 & 8 & 9 \\ \hline & 3 & 3 & 6 \end{array}$$

Review Video: Subtracting Large Numbers
Visit mometrix.com/academy and enter code: 603350

Order of Operations

The **order of operations** is a set of rules that dictates the order in which we must perform each operation in an expression so that we will evaluate it accurately. If we have an expression that includes multiple different operations, the order of operations tells us which operations to do first. The most common mnemonic for the order of operations is **PEMDAS**, or "Please Excuse My Dear Aunt Sally." PEMDAS stands for parentheses, exponents, multiplication, division, addition, and subtraction. It is important to understand that multiplication and division have equal precedence, as do addition and subtraction, so those pairs of operations are simply worked from left to right in order.

Copyright © Mometrix Media. You have been licensed one copy of this document for personal use only. Any other reproduction or redistribution is strictly prohibited. All rights reserved.
This content is provided for test preparation purposes only and does not imply an endorsement by Mometrix of any particular political, scientific, or religious point of view.

For example, evaluating the expression $5 + 20 \div 4 \times (2 + 3)^2 - 6$ using the correct order of operations would be done like this:

- **P:** Perform the operations inside the parentheses: $(2 + 3) = 5$
- **E:** Simplify the exponents: $(5)^2 = 5 \times 5 = 25$
  - The expression now looks like this: $5 + 20 \div 4 \times 25 - 6$
- **MD:** Perform multiplication and division from left to right: $20 \div 4 = 5$; then $5 \times 25 = 125$
  - The expression now looks like this: $5 + 125 - 6$
- **AS:** Perform addition and subtraction from left to right: $5 + 125 = 130$; then $130 - 6 = 124$

Review Video: Order of Operations
Visit mometrix.com/academy and enter code: 259675

Properties of Exponents

The properties of exponents are as follows:

Property	Description
$a^1 = a$	Any number to the power of 1 is equal to itself
$1^n = 1$	The number 1 raised to any power is equal to 1
$a^0 = 1$	Any number raised to the power of 0 is equal to 1
$a^n \times a^m = a^{n+m}$	Add exponents to multiply powers of the same base number
$a^n \div a^m = a^{n-m}$	Subtract exponents to divide powers of the same base number
$(a^n)^m = a^{n\times m}$	When a power is raised to a power, the exponents are multiplied
$(a \times b)^n = a^n \times b^n$ $(a \div b)^n = a^n \div b^n$	Multiplication and division operations inside parentheses can be raised to a power. This is the same as each term being raised to that power.
$a^{-n} = \frac{1}{a^n}$	A negative exponent is the same as the reciprocal of a positive exponent

Note that exponents do not have to be integers. Fractional or decimal exponents follow all the rules above as well. Example: $5^{\frac{1}{4}} \times 5^{\frac{3}{4}} = 5^{\frac{1}{4}+\frac{3}{4}} = 5^1 = 5$.

Review Video: Properties of Exponents
Visit mometrix.com/academy and enter code: 532558

Scientific Notation

Scientific notation is a way of writing large numbers in a shorter form. The form $a \times 10^n$ is used in scientific notation, where a is greater than or equal to 1 but less than 10, and n is the number of places the decimal must move to get from the original number to a. Example: The number 230,400,000 is cumbersome to write. To write the value in scientific notation, place a decimal point between the first and second numbers, and include all digits through the last non-zero digit ($a = 2.304$). To find the appropriate power of 10, count the number of places the decimal point had to move ($n = 8$). The number is positive if the decimal moved to the left, and negative if it moved to the right. We can then write 230,400,000 as 2.304×10^8. If we look instead at the number 0.00002304, we have the same value for a, but this time the decimal moved 5 places to the right

Copyright © Mometrix Media. You have been licensed one copy of this document for personal use only. Any other reproduction or redistribution is strictly prohibited. All rights reserved.
This content is provided for test preparation purposes only and does not imply an endorsement by Mometrix of any particular political, scientific, or religious point of view.

($n = -5$). Thus, 0.00002304 can be written as 2.304×10^{-5}. Using this notation makes it simple to compare very large or very small numbers. By comparing exponents, it is easy to see that 3.28×10^4 is smaller than 1.51×10^5, because 4 is less than 5.

Review Video: Scientific Notation
Visit mometrix.com/academy and enter code: 976454

Factors and Multiples

Factors and Greatest Common Factor

Factors are numbers that are multiplied together to obtain a **product**. For example, in the equation $2 \times 3 = 6$, the numbers 2 and 3 are factors. A **prime number** has only two factors (1 and itself), but other numbers can have many factors.

A **common factor** is a number that divides exactly into two or more other numbers. For example, the factors of 12 are 1, 2, 3, 4, 6, and 12, while the factors of 15 are 1, 3, 5, and 15. The common factors of 12 and 15 are 1 and 3.

A **prime factor** is also a prime number. Therefore, the prime factors of 12 are 2 and 3. For 15, the prime factors are 3 and 5.

The **greatest common factor** (GCF) is the largest number that is a factor of two or more numbers. For example, the factors of 15 are 1, 3, 5, and 15; the factors of 35 are 1, 5, 7, and 35. Therefore, the greatest common factor of 15 and 35 is 5.

Review Video: Factors
Visit mometrix.com/academy and enter code: 920086

Review Video: Prime Numbers and Factorization
Visit mometrix.com/academy and enter code: 760669

Review Video: Greatest Common Factor and Least Common Multiple
Visit mometrix.com/academy and enter code: 838699

Multiples and Least Common Multiple

Often listed out in multiplication tables, **multiples** are integer increments of a given factor. In other words, dividing a multiple by the factor will result in an integer. For example, the multiples of 7 include: $1 \times 7 = 7$, $2 \times 7 = 14$, $3 \times 7 = 21$, $4 \times 7 = 28$, $5 \times 7 = 35$. Dividing 7, 14, 21, 28, or 35 by 7 will result in the integers 1, 2, 3, 4, and 5, respectively.

The least common multiple (**LCM**) is the smallest number that is a multiple of two or more numbers. For example, the multiples of 3 include 3, 6, 9, 12, 15, etc.; the multiples of 5 include 5, 10, 15, 20, etc. Therefore, the least common multiple of 3 and 5 is 15.

Review Video: Multiples
Visit mometrix.com/academy and enter code: 626738

Copyright © Mometrix Media. You have been licensed one copy of this document for personal use only. Any other reproduction or redistribution is strictly prohibited. All rights reserved.
This content is provided for test preparation purposes only and does not imply an endorsement by Mometrix of any particular political, scientific, or religious point of view.

Fractions, Decimals, and Percentages

Fractions

A **fraction** is a number that is expressed as one integer written above another integer, with a dividing line between them $\left(\frac{x}{y}\right)$. It represents the **quotient** of the two numbers "x divided by y." It can also be thought of as x out of y equal parts.

The top number of a fraction is called the **numerator**, and it represents the number of parts under consideration. The 1 in $\frac{1}{4}$ means that 1 part out of the whole is being considered in the calculation. The bottom number of a fraction is called the **denominator**, and it represents the total number of equal parts. The 4 in $\frac{1}{4}$ means that the whole consists of 4 equal parts. A fraction cannot have a denominator of zero; this is referred to as "*undefined*."

Fractions can be manipulated, without changing the value of the fraction, by multiplying or dividing (but not adding or subtracting) both the numerator and denominator by the same number. If you divide both numbers by a common factor, you are **reducing** or simplifying the fraction. Two fractions that have the same value but are expressed differently are known as **equivalent fractions**. For example, $\frac{2}{10}, \frac{3}{15}, \frac{4}{20}$, and $\frac{5}{25}$ are all equivalent fractions. They can also all be reduced or simplified to $\frac{1}{5}$.

When two fractions are manipulated so that they have the same denominator, this is known as finding a **common denominator**. The number chosen to be that common denominator should be the least common multiple of the two original denominators. Example: $\frac{3}{4}$ and $\frac{5}{6}$; the least common multiple of 4 and 6 is 12. Manipulating to achieve the common denominator: $\frac{3}{4} = \frac{9}{12}; \frac{5}{6} = \frac{10}{12}$.

Review Video: Overview of Fractions
Visit mometrix.com/academy and enter code: 262335

Proper Fractions and Mixed Numbers

A fraction whose denominator is greater than its numerator is known as a **proper fraction**, while a fraction whose numerator is greater than its denominator is known as an **improper fraction**. Proper fractions have values *less than one* and improper fractions have values *greater than one.*

A **mixed number** is a number that contains both an integer and a fraction. Any improper fraction can be rewritten as a mixed number. Example: $\frac{8}{3} = \frac{6}{3} + \frac{2}{3} = 2 + \frac{2}{3} = 2\frac{2}{3}$. Similarly, any mixed number can be rewritten as an improper fraction. Example: $1\frac{3}{5} = 1 + \frac{3}{5} = \frac{5}{5} + \frac{3}{5} = \frac{8}{5}$.

Review Video: Proper and Improper Fractions and Mixed Numbers
Visit mometrix.com/academy and enter code: 211077

Copyright © Mometrix Media. You have been licensed one copy of this document for personal use only. Any other reproduction or redistribution is strictly prohibited. All rights reserved.
This content is provided for test preparation purposes only and does not imply an endorsement by Mometrix of any particular political, scientific, or religious point of view.

Adding and Subtracting Fractions

If two fractions have a common denominator, they can be added or subtracted simply by adding or subtracting the two numerators and retaining the same denominator. If the two fractions do not already have the same denominator, one or both of them must be manipulated to achieve a common denominator before they can be added or subtracted. Example: $\frac{1}{2} + \frac{1}{4} = \frac{2}{4} + \frac{1}{4} = \frac{3}{4}$.

Review Video: Adding and Subtracting Fractions
Visit mometrix.com/academy and enter code: 378080

Multiplying Fractions

Two fractions can be multiplied by multiplying the two numerators to find the new numerator and the two denominators to find the new denominator. Example: $\frac{1}{3} \times \frac{2}{3} = \frac{1\times2}{3\times3} = \frac{2}{9}$.

Dividing Fractions

Two fractions can be divided by flipping the numerator and denominator of the second fraction and then proceeding as though it were a multiplication problem. Example: $\frac{2}{3} \div \frac{3}{4} = \frac{2}{3} \times \frac{4}{3} = \frac{8}{9}$.

Review Video: Multiplying and Dividing Fractions
Visit mometrix.com/academy and enter code: 473632

Multiplying a Mixed Number by a Whole Number or a Decimal

When multiplying a mixed number by something, it is usually best to convert it to an improper fraction first. Additionally, if the multiplicand is a decimal, it is most often simplest to convert it to a fraction. For instance, to multiply $4\frac{3}{8}$ by 3.5, begin by rewriting each quantity as a whole number plus a proper fraction. Remember, a mixed number is a fraction added to a whole number and a decimal is a representation of the sum of fractions, specifically tenths, hundredths, thousandths, and so on:

$$4\frac{3}{8} \times 3.5 = \left(4 + \frac{3}{8}\right) \times \left(3 + \frac{1}{2}\right)$$

Next, the quantities being added need to be expressed with the same denominator. This is achieved by multiplying and dividing the whole number by the denominator of the fraction. Recall that a whole number is equivalent to that number divided by 1:

$$= \left(\frac{4}{1} \times \frac{8}{8} + \frac{3}{8}\right) \times \left(\frac{3}{1} \times \frac{2}{2} + \frac{1}{2}\right)$$

When multiplying fractions, remember to multiply the numerators and denominators separately:

$$= \left(\frac{4 \times 8}{1 \times 8} + \frac{3}{8}\right) \times \left(\frac{3 \times 2}{1 \times 2} + \frac{1}{2}\right)$$
$$= \left(\frac{32}{8} + \frac{3}{8}\right) \times \left(\frac{6}{2} + \frac{1}{2}\right)$$

Now that the fractions have the same denominators, they can be added:

$$= \frac{35}{8} \times \frac{7}{2}$$

Copyright © Mometrix Media. You have been licensed one copy of this document for personal use only. Any other reproduction or redistribution is strictly prohibited. All rights reserved.
This content is provided for test preparation purposes only and does not imply an endorsement by Mometrix of any particular political, scientific, or religious point of view.

Finally, perform the last multiplication and then simplify:

$$= \frac{35 \times 7}{8 \times 2} = \frac{245}{16} = \frac{240}{16} + \frac{5}{16} = 15\frac{5}{16}$$

Comparing Fractions

It is important to master the ability to compare and order fractions. This skill is relevant to many real-world scenarios. For example, carpenters often compare fractional construction nail lengths when preparing for a project, and bakers often compare fractional measurements to have the correct ratio of ingredients. There are three commonly used strategies when comparing fractions. These strategies are referred to as the common denominator approach, the decimal approach, and the cross-multiplication approach.

Using a Common Denominator to Compare Fractions

The fractions $\frac{2}{3}$ and $\frac{4}{7}$ have different denominators. $\frac{2}{3}$ has a denominator of 3, and $\frac{4}{7}$ has a denominator of 7. In order to precisely compare these two fractions, it is necessary to use a common denominator. A common denominator is a common multiple that is shared by both denominators. In this case, the denominators 3 and 7 share a multiple of 21. In general, it is most efficient to select the least common multiple for the two denominators.

Rewrite each fraction with the common denominator of 21. Then, calculate the new numerators as illustrated below.

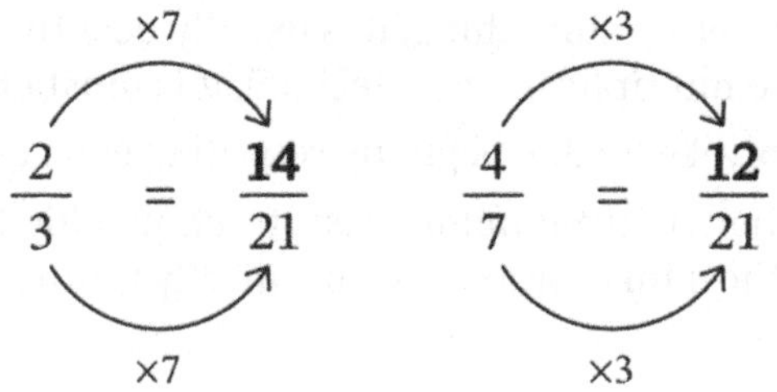

For $\frac{2}{3}$, multiply the numerator and denominator by 7. The result is $\frac{14}{21}$.

For $\frac{4}{7}$, multiply the numerator and denominator by 3. The result is $\frac{12}{21}$.

Now that both fractions have a denominator of 21, the fractions can accurately be compared by comparing the numerators. Since 14 is greater than 12, the fraction $\frac{14}{21}$ is greater than $\frac{12}{21}$. This means that $\frac{2}{3}$ is greater than $\frac{4}{7}$.

Using Decimals to Compare Fractions

Sometimes decimal values are easier to compare than fraction values. For example, $\frac{5}{8}$ is equivalent to 0.625 and $\frac{3}{5}$ is equivalent to 0.6. This means that the comparison of $\frac{5}{8}$ and $\frac{3}{5}$ can be determined by comparing the decimals 0.625 and 0.6. When both decimal values are extended to the thousandths place, they become 0.625 and 0.600, respectively. It becomes clear that 0.625 is greater than 0.600 because 625 thousandths is greater than 600 thousandths. In other words, $\frac{5}{8}$ is greater than $\frac{3}{5}$ because 0.625 is greater than 0.6.

Copyright © Mometrix Media. You have been licensed one copy of this document for personal use only. Any other reproduction or redistribution is strictly prohibited. All rights reserved.
This content is provided for test preparation purposes only and does not imply an endorsement by Mometrix of any particular political, scientific, or religious point of view.

Using Cross-Multiplication to Compare Fractions

Cross-multiplication is an efficient strategy for comparing fractions. This is a shortcut for the common denominator strategy. Start by writing each fraction next to one another. Multiply the numerator of the fraction on the left by the denominator of the fraction on the right. Write down the result next to the fraction on the left. Now multiply the numerator of the fraction on the right by the denominator of the fraction on the left. Write down the result next to the fraction on the right. Compare both products. The fraction with the larger result is the larger fraction.

Consider the fractions $\frac{4}{7}$ and $\frac{5}{9}$.

36 is greater than 35. Therefore, $\frac{4}{7}$ is greater than $\frac{5}{9}$.

Decimals

Decimals are one way to represent parts of a whole. Using the place value system, each digit to the right of a decimal point denotes the number of units of a corresponding *negative* power of ten. For example, consider the decimal 0.24. We can use a model to represent the decimal. Since a dime is worth one-tenth of a dollar and a penny is worth one-hundredth of a dollar, one possible model to represent this fraction is to have 2 dimes representing the 2 in the tenths place and 4 pennies representing the 4 in the hundredths place:

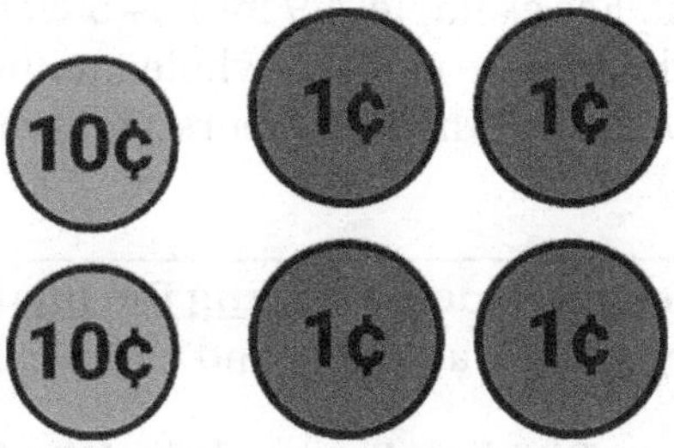

To write the decimal as a fraction, put the decimal in the numerator with 1 in the denominator. Multiply the numerator and denominator by tens until there are no more decimal places. Then simplify the fraction to lowest terms. For example, converting 0.24 to a fraction:

$$0.24 = \frac{0.24}{1} = \frac{0.24 \times 100}{1 \times 100} = \frac{24}{100} = \frac{6}{25}$$

Review Video: Decimals
Visit mometrix.com/academy and enter code: 837268

Operations with Decimals

Adding and Subtracting Decimals

When adding and subtracting decimals, the decimal points must always be aligned. Adding decimals is just like adding regular whole numbers. Example: $4.5 + 2.0 = 6.5$.

Copyright © Mometrix Media. You have been licensed one copy of this document for personal use only. Any other reproduction or redistribution is strictly prohibited. All rights reserved.
This content is provided for test preparation purposes only and does not imply an endorsement by Mometrix of any particular political, scientific, or religious point of view.

If the problem-solver does not properly align the decimal points, an incorrect answer of 4.7 may result. An easy way to add decimals is to align all of the decimal points in a vertical column visually. This will allow you to see exactly where the decimal should be placed in the final answer. Begin adding from right to left. Add each column in turn, making sure to carry the number to the left if a column adds up to more than 9. The same rules apply to the subtraction of decimals.

Review Video: Adding and Subtracting Decimals
Visit mometrix.com/academy and enter code: 381101

Multiplying Decimals

A simple multiplication problem has two components: a **multiplicand** and a **multiplier**. When multiplying decimals, work as though the numbers were whole rather than decimals. Once the final product is calculated, count the number of places to the right of the decimal in both the multiplicand and the multiplier. Then, count that number of places from the right of the product and place the decimal in that position.

For example, 12.3×2.56 has a total of three places to the right of the respective decimals. Multiply 123×256 to get 31,488. Now, beginning on the right, count three places to the left and insert the decimal. The final product will be 31.488.

Review Video: How to Multiply Decimals
Visit mometrix.com/academy and enter code: 731574

Dividing Decimals

Every division problem has a **divisor** and a **dividend**. The dividend is the number that is being divided. In the problem $14 \div 7$, 14 is the dividend and 7 is the divisor. In a division problem with decimals, the divisor must be converted into a whole number. Begin by moving the decimal in the divisor to the right until a whole number is created. Next, move the decimal in the dividend the same number of spaces to the right. For example, 4.9 into 24.5 would become 49 into 245. The decimal was moved one space to the right to create a whole number in the divisor, and then the same was done for the dividend. Once the whole numbers are created, the problem is carried out normally: $245 \div 49 = 5$.

Review Video: Dividing Decimals
Visit mometrix.com/academy and enter code: 560690

Review Video: Dividing Decimals by Whole Numbers
Visit mometrix.com/academy and enter code: 535669

Percentages

Percentages can be thought of as fractions that are based on a whole of 100; that is, one whole is equal to 100%. The word **percent** means "per hundred." Percentage problems are often presented in three main ways:

- Find what percentage of some number another number is.
  - Example: What percentage of 40 is 8?
- Find what number is some percentage of a given number.
  - Example: What number is 20% of 40?

Copyright © Mometrix Media. You have been licensed one copy of this document for personal use only. Any other reproduction or redistribution is strictly prohibited. All rights reserved.
This content is provided for test preparation purposes only and does not imply an endorsement by Mometrix of any particular political, scientific, or religious point of view.

- Find what number another number is a given percentage of.
  - Example: What number is 8 20% of?

There are three components in each of these cases: a **whole** (W), a **part** (P), and a **percentage** (%). These are related by the equation: $P = W \times \%$. This can easily be rearranged into other forms that may suit different questions better: $\% = \frac{P}{W}$ and $W = \frac{P}{\%}$. Percentage problems are often also word problems. As such, a large part of solving them is figuring out which quantities are what. For example, consider the following word problem:

In a school cafeteria, 7 students choose pizza, 9 choose hamburgers, and 4 choose tacos. What percentage of student choose tacos?

To find the whole, you must first add all of the parts: $7 + 9 + 4 = 20$. The percentage can then be found by dividing the part by the whole $\left(\% = \frac{P}{W}\right)$: $\frac{4}{20} = \frac{20}{100} = 20\%$.

Review Video: Computation with Percentages
Visit mometrix.com/academy and enter code: 693099

Converting Between Percentages, Fractions, and Decimals

Converting decimals to percentages and percentages to decimals is as simple as moving the decimal point. To *convert from a decimal to a percentage*, move the decimal point **two places to the right**. To *convert from a percentage to a decimal*, move it **two places to the left**. It may be helpful to remember that the percentage number will always be larger than the equivalent decimal number. Example:

$$0.23 = 23\% \quad 5.34 = 534\% \quad 0.007 = 0.7\%$$
$$700\% = 7.00 \quad 86\% = 0.86 \quad 0.15\% = 0.0015$$

To convert a fraction to a decimal, simply divide the numerator by the denominator in the fraction. To convert a decimal to a fraction, put the decimal in the numerator with 1 in the denominator. Multiply the numerator and denominator by tens until there are no more decimal places. Then simplify the fraction to lowest terms. For example, converting 0.24 to a fraction:

$$0.24 = \frac{0.24}{1} = \frac{0.24 \times 100}{1 \times 100} = \frac{24}{100} = \frac{6}{25}$$

Fractions can be converted to a percentage by finding equivalent fractions with a denominator of 100. Example:

$$\frac{7}{10} = \frac{70}{100} = 70\% \quad \frac{1}{4} = \frac{25}{100} = 25\%$$

To convert a percentage to a fraction, divide the percentage number by 100 and reduce the fraction to its simplest possible terms. Example:

$$60\% = \frac{60}{100} = \frac{3}{5} \quad 96\% = \frac{96}{100} = \frac{24}{25}$$

Review Video: Converting Fractions to Percentages and Decimals
Visit mometrix.com/academy and enter code: 306233

Copyright © Mometrix Media. You have been licensed one copy of this document for personal use only. Any other reproduction or redistribution is strictly prohibited. All rights reserved.
This content is provided for test preparation purposes only and does not imply an endorsement by Mometrix of any particular political, scientific, or religious point of view.

Review Video: Converting Percentages to Decimals and Fractions
Visit mometrix.com/academy and enter code: 287297

Review Video: Converting Decimals to Fractions and Percentages
Visit mometrix.com/academy and enter code: 986765

Review Video: Converting Decimals, Improper Fractions, and Mixed Numbers
Visit mometrix.com/academy and enter code: 696924

Proportions and Ratios

PROPORTIONS

A proportion is a relationship between two quantities that dictates how one changes when the other changes. A **direct proportion** describes a relationship in which a quantity increases by a set amount for every increase in the other quantity, or decreases by that same amount for every decrease in the other quantity. Example: Assuming a constant driving speed, the time required for a car trip increases as the distance of the trip increases. The distance to be traveled and the time required to travel are directly proportional.

An **inverse proportion** is a relationship in which an increase in one quantity is accompanied by a decrease in the other, or vice versa. Example: the time required for a car trip decreases as the speed increases and increases as the speed decreases, so the time required is inversely proportional to the speed of the car.

Review Video: Proportions
Visit mometrix.com/academy and enter code: 505355

RATIOS

A **ratio** is a comparison of two quantities in a particular order. Example: If there are 14 computers in a lab, and the class has 20 students, there is a student to computer ratio of 20 to 14, commonly written as 20: 14. Ratios are normally reduced to their smallest whole number representation, so 20: 14 would be reduced to 10: 7 by dividing both sides by 2.

Review Video: Ratios
Visit mometrix.com/academy and enter code: 996914

CONSTANT OF PROPORTIONALITY

When two quantities have a proportional relationship, there exists a **constant of proportionality** between the quantities. The product of this constant and one of the quantities is equal to the other quantity. For example, if one lemon costs $0.25, two lemons cost $0.50, and three lemons cost $0.75, there is a proportional relationship between the total cost of lemons and the number of lemons purchased. The constant of proportionality is the **unit price**, namely $0.25/lemon. Notice that the total price of lemons, t, can be found by multiplying the unit price of lemons, p, and the number of lemons, n: $t = pn$.

WORK/UNIT RATE

Unit rate expresses a quantity of one thing in terms of one unit of another. For example, if you travel 30 miles every two hours, a unit rate expresses this comparison in terms of one hour: in one hour you travel 15 miles, so your unit rate is 15 miles per hour. Other examples are how much one ounce of food costs (price per ounce) or figuring out how much one egg costs out of the dozen

Copyright © Mometrix Media. You have been licensed one copy of this document for personal use only. Any other reproduction or redistribution is strictly prohibited. All rights reserved.
This content is provided for test preparation purposes only and does not imply an endorsement by Mometrix of any particular political, scientific, or religious point of view.

(price per 1 egg, instead of price per 12 eggs). The denominator of a unit rate is always 1. Unit rates are used to compare different situations to solve problems. For example, to make sure you get the best deal when deciding which kind of soda to buy, you can find the unit rate of each. If soda #1 costs $1.50 for a 1-liter bottle, and soda #2 costs $2.75 for a 2-liter bottle, it would be a better deal to buy soda #2, because its unit rate is only $1.375 per 1-liter, which is cheaper than soda #1. Unit rates can also help determine the length of time a given event will take. For example, if you can paint 2 rooms in 4.5 hours, you can determine how long it will take you to paint 5 rooms by solving for the unit rate per room and then multiplying that by 5.

> **Review Video:** Rates and Unit Rates
> Visit mometrix.com/academy and enter code: 185363

Linear Expressions

Terms and Coefficients

Mathematical expressions consist of a combination of one or more values arranged in terms that are added together. As such, an expression could be just a single number, including zero. A **variable term** is the product of a real number, also called a **coefficient**, and one or more variables, each of which may be raised to an exponent. Expressions may also include numbers without a variable, called **constants** or **constant terms**. The expression $6s^2$, for example, is a single term where the coefficient is the real number 6 and the variable term is s^2. Note that if a term is written as simply a variable to some exponent, like t^2, then the coefficient is 1, because $t^2 = 1t^2$.

Linear Expressions

A **single variable linear expression** is the sum of a single variable term, where the variable has no exponent, and a constant, which may be zero. For instance, the expression $2w + 7$ has $2w$ as the variable term and 7 as the constant term. It is important to realize that terms are separated by addition or subtraction. Since an expression is a sum of terms, expressions such as $5x - 3$ can be written as $5x + (-3)$ to emphasize that the constant term is negative. A real-world example of a single variable linear expression is the perimeter of a square, four times the side length, often expressed: $4s$.

In general, a **linear expression** is the sum of any number of variable terms so long as none of the variables have an exponent. For example, $3m + 8n - \frac{1}{4}p + 5.5q - 1$ is a linear expression, but $3y^3$ is not. In the same way, the expression for the perimeter of a general triangle, the sum of the side lengths $(a + b + c)$ is considered to be linear, but the expression for the area of a square, the side length squared (s^2) is not.

Slope

On a graph with two points, (x_1, y_1) and (x_2, y_2), the **slope** is found with the formula $m = \frac{y_2 - y_1}{x_2 - x_1}$; where $x_1 \neq x_2$ and m stands for slope. If the value of the slope is **positive**, the line has an *upward direction* from left to right. If the value of the slope is **negative**, the line has a *downward direction* from left to right. Consider the following example:

Copyright © Mometrix Media. You have been licensed one copy of this document for personal use only. Any other reproduction or redistribution is strictly prohibited. All rights reserved.
This content is provided for test preparation purposes only and does not imply an endorsement by Mometrix of any particular political, scientific, or religious point of view.

A new book goes on sale in bookstores and online stores. In the first month, 5,000 copies of the book are sold. Over time, the book continues to grow in popularity. The data for the number of copies sold is in the table below.

# of Months on Sale	1	2	3	4	5
# of Copies Sold (In Thousands)	5	10	15	20	25

So, the number of copies that are sold and the time that the book is on sale is a proportional relationship. In this example, an equation can be used to show the data: $y = 5x$, where x is the number of months that the book is on sale. Also, y is the number of copies sold. So, the slope of the corresponding line is $\frac{\text{rise}}{\text{run}} = \frac{5}{1} = 5$.

Review Video: Finding the Slope of a Line
Visit mometrix.com/academy and enter code: 766664

Linear Equations

Equations that can be written as $ax + b = 0$, where $a \neq 0$, are referred to as **one variable linear equations.** A solution to such an equation is called a **root**. In the case where we have the equation $5x + 10 = 0$, if we solve for x we get a solution of $x = -2$. In other words, the root of the equation is -2. This is found by first subtracting 10 from both sides, which gives $5x = -10$. Next, simply divide both sides by the coefficient of the variable, in this case 5, to get $x = -2$. This can be checked by plugging -2 back into the original equation $(5)(-2) + 10 = -10 + 10 = 0$.

The **solution set** is the set of all solutions of an equation. In our example, the solution set would simply be -2. If there were more solutions (there usually are in multivariable equations) then they would also be included in the solution set. When an equation has no true solutions, it is referred to as an **empty set**. Equations with identical solution sets are **equivalent equations**. An **identity** is a term whose value or determinant is equal to 1.

Linear equations can be written many ways. Below is a list of some forms linear equations can take:

- **Standard Form**: $Ax + By = C$; the slope is $\frac{-A}{B}$ and the y-intercept is $\frac{C}{B}$
- **Slope Intercept Form**: $y = mx + b$, where m is the slope and b is the y-intercept
- **Point-Slope Form**: $y - y_1 = m(x - x_1)$, where m is the slope and (x_1, y_1) is a point on the line
- **Two-Point Form**: $\frac{y-y_1}{x-x_1} = \frac{y_2-y_1}{x_2-x_1}$, where (x_1, y_1) and (x_2, y_2) are two points on the given line
- **Intercept Form**: $\frac{x}{x_1} + \frac{y}{y_1} = 1$, where $(x_1, 0)$ is the point at which a line intersects the x-axis, and $(0, y_1)$ is the point at which the same line intersects the y-axis

Review Video: Slope-Intercept and Point-Slope Forms
Visit mometrix.com/academy and enter code: 113216

Review Video: Linear Equations Basics
Visit mometrix.com/academy and enter code: 793005

Copyright © Mometrix Media. You have been licensed one copy of this document for personal use only. Any other reproduction or redistribution is strictly prohibited. All rights reserved.
This content is provided for test preparation purposes only and does not imply an endorsement by Mometrix of any particular political, scientific, or religious point of view.

Solving Equations

Solving One-Variable Linear Equations

Multiply all terms by the lowest common denominator to eliminate any fractions. Look for addition or subtraction to undo so you can isolate the variable on one side of the equal sign. Divide both sides by the coefficient of the variable. When you have a value for the variable, substitute this value into the original equation to make sure you have a true equation. Consider the following example:

Kim's savings are represented by the table below. Represent her savings, using an equation.

X (Months)	Y (Total Savings)
2	$1,300
5	$2,050
9	$3,050
11	$3,550
16	$4,800

The table shows a function with a constant rate of change, or slope, of 250. Given the points on the table, the slopes can be calculated as $\frac{(2{,}050-1300)}{(5-2)}$, $\frac{(3{,}050-2{,}050)}{(9-5)}$, $\frac{(3{,}550-3{,}050)}{(11-9)}$, and $\frac{(4{,}800-3{,}550)}{(16-11)}$, each of which equals 250. Thus, the table shows a constant rate of change, indicating a linear function. The slope-intercept form of a linear equation is written as $y = mx + b$, where m represents the slope and b represents the y-intercept. Substituting the slope into this form gives $y = 250x + b$. Substituting corresponding x- and y-values from any point into this equation will give the y-intercept, or b. Using the point, $(2, 1{,}300)$, gives $1{,}300 = 250(2) + b$, which simplifies as $b = 800$. Thus, her savings may be represented by the equation, $y = 250x + 800$.

Rules for Manipulating Equations

Like Terms

Like terms are terms in an equation that have the same variable, regardless of whether or not they also have the same coefficient. This includes terms that *lack* a variable; all constants (i.e., numbers without variables) are considered like terms. If the equation involves terms with a variable raised to different powers, the like terms are those that have the variable raised to the same power.

For example, consider the equation $x^2 + 3x + 2 = 2x^2 + x - 7 + 2x$. In this equation, 2 and -7 are like terms; they are both constants. $3x$, x, and $2x$ are like terms, they all include the variable x raised to the first power. x^2 and $2x^2$ are like terms, they both include the variable x, raised to the second power. $2x$ and $2x^2$ are not like terms; although they both involve the variable x, the variable is not raised to the same power in both terms. The fact that they have the same coefficient, 2, is not relevant.

Review Video: Rules for Manipulating Equations
Visit mometrix.com/academy and enter code: 838871

Carrying Out the Same Operation on Both Sides of an Equation

When solving an equation, the general procedure is to carry out a series of operations on both sides of an equation, choosing operations that will tend to simplify the equation when doing so. The reason why the same operation must be carried out on both sides of the equation is because that

Copyright © Mometrix Media. You have been licensed one copy of this document for personal use only. Any other reproduction or redistribution is strictly prohibited. All rights reserved.
This content is provided for test preparation purposes only and does not imply an endorsement by Mometrix of any particular political, scientific, or religious point of view.

leaves the meaning of the equation unchanged, and yields a result that is equivalent to the original equation. This would not be the case if we carried out an operation on one side of an equation and not the other. Consider what an equation means: it is a statement that two values or expressions are equal. If we carry out the same operation on both sides of the equation—add 3 to both sides, for example—then the two sides of the equation are changed in the same way, and so remain equal. If we do that to only one side of the equation—add 3 to one side but not the other—then that wouldn't be true; if we change one side of the equation but not the other then the two sides are no longer equal.

Advantage of Combining Like Terms

Combining like terms refers to adding or subtracting like terms—terms with the same variable—and therefore reducing sets of like terms to a single term. The main advantage of doing this is that it simplifies the equation. Often, combining like terms can be done as the first step in solving an equation, though it can also be done later, such as after distributing terms in a product.

For example, consider the equation $2(x + 3) + 3(2 + x + 3) = -4$. The 2 and the 3 in the second set of parentheses are like terms, and we can combine them, yielding $2(x + 3) + 3(x + 5) = -4$. Now we can carry out the multiplications implied by the parentheses, distributing the outer 2 and 3 accordingly: $2x + 6 + 3x + 15 = -4$. The $2x$ and the $3x$ are like terms, and we can add them together: $5x + 6 + 15 = -4$. Now, the constants 6, 15, and -4 are also like terms, and we can combine them as well: subtracting 6 and 15 from both sides of the equation, we get $5x = -4 - 6 - 15$, or $5x = -25$, which simplifies further to $x = -5$.

Review Video: Solving Equations by Combining Like Terms
Visit mometrix.com/academy and enter code: 668506

Canceling Terms on Opposite Sides of an Equation

Two terms on opposite sides of an equation can be canceled if and only if they *exactly* match each other. They must have the same variable raised to the same power and the same coefficient. For example, in the equation $3x + 2x^2 + 6 = 2x^2 - 6$, $2x^2$ appears on both sides of the equation and can be canceled, leaving $3x + 6 = -6$. The 6 on each side of the equation *cannot* be canceled, because it is added on one side of the equation and subtracted on the other. While they cannot be canceled, however, the 6 and -6 are like terms and can be combined, yielding $3x = -12$, which simplifies further to $x = -4$.

It's also important to note that the terms to be canceled must be independent terms and cannot be part of a larger term. For example, consider the equation $2(x + 6) = 3(x + 4) + 1$. We cannot cancel the x's, because even though they match each other they are part of the larger terms $2(x + 6)$ and $3(x + 4)$. We must first distribute the 2 and 3, yielding $2x + 12 = 3x + 12 + 1$. Now we see that the terms with the x's do not match, but the 12s do, and can be canceled, leaving $2x = 3x + 1$, which simplifies to $x = -1$.

Process for Manipulating Equations

Isolating Variables

To **isolate a variable** means to manipulate the equation so that the variable appears by itself on one side of the equation, and does not appear at all on the other side. Generally, an equation or inequality is considered to be solved once the variable is isolated and the other side of the equation or inequality is simplified as much as possible. In the case of a two-variable equation or inequality, only one variable needs to be isolated; it will not usually be possible to simultaneously isolate both variables.

Copyright © Mometrix Media. You have been licensed one copy of this document for personal use only. Any other reproduction or redistribution is strictly prohibited. All rights reserved.
This content is provided for test preparation purposes only and does not imply an endorsement by Mometrix of any particular political, scientific, or religious point of view.

For a linear equation—an equation in which the variable only appears raised to the first power—isolating a variable can be done by first moving all the terms with the variable to one side of the equation and all other terms to the other side. (*Moving* a term really means adding the inverse of the term to both sides; when a term is *moved* to the other side of the equation its sign is flipped.) Then combine like terms on each side. Finally, divide both sides by the coefficient of the variable, if applicable. The steps need not necessarily be done in this order, but this order will always work.

Review Video: Solving One-Step Equations
Visit mometrix.com/academy and enter code: 777004

Equations with More Than One Solution

Some types of non-linear equations, such as equations involving squares of variables, may have more than one solution. For example, the equation $x^2 = 4$ has two solutions: 2 and –2. Equations with absolute values can also have multiple solutions: $|x| = 1$ has the solutions $x = 1$ and $x = -1$.

It is also possible for a linear equation to have more than one solution, but only if the equation is true regardless of the value of the variable. In this case, the equation is considered to have infinitely many solutions, because any possible value of the variable is a solution. We know a linear equation has infinitely many solutions if when we combine like terms the variables cancel, leaving a true statement. For example, consider the equation $2(3x + 5) = x + 5(x + 2)$. Distributing, we get $6x + 10 = x + 5x + 10$; combining like terms gives $6x + 10 = 6x + 10$, and the $6x$-terms cancel to leave $10 = 10$. This is clearly true, so the original equation is true for any value of x. We could also have canceled the 10s leaving $0 = 0$, but again this is clearly true—in general if both sides of the equation match exactly, it has infinitely many solutions.

Equations with No Solution

Some types of non-linear equations, such as equations involving squares of variables, may have no solution. For example, the equation $x^2 = -2$ has no solutions in the real numbers, because the square of any real number must be positive. Similarly, $|x| = -1$ has no solution, because the absolute value of a number is always positive.

It is also possible for an equation to have no solution even if does not involve any powers greater than one, absolute values, or other special functions. For example, the equation $2(x + 3) + x = 3x$ has no solution. We can see that if we try to solve it: first we distribute, leaving $2x + 6 + x = 3x$. But now if we try to combine all the terms with the variable, we find that they cancel: we have $3x$ on the left and $3x$ on the right, canceling to leave us with $6 = 0$. This is clearly false. In general, whenever the variable terms in an equation cancel leaving different constants on both sides, it means that the equation has no solution. (If we are left with the *same* constant on both sides, the equation has infinitely many solutions instead.)

Features of Equations That Require Special Treatment

Linear Equations

A linear equation is an equation in which variables only appear by themselves: not multiplied together, not with exponents other than one, and not inside absolute value signs or any other functions. For example, the equation $x + 1 - 3x = 5 - x$ is a linear equation; while x appears multiple times, it never appears with an exponent other than one, or inside any function. The two-variable equation $2x - 3y = 5 + 2x$ is also a linear equation. In contrast, the equation $x^2 - 5 = 3x$ is *not* a linear equation, because it involves the term x^2. $\sqrt{x} = 5$ is not a linear equation, because it involves a square root. $(x - 1)^2 = 4$ is not a linear equation because even though there's no exponent on the x directly, it appears as part of an expression that is squared. The two-variable

Copyright © Mometrix Media. You have been licensed one copy of this document for personal use only. Any other reproduction or redistribution is strictly prohibited. All rights reserved.
This content is provided for test preparation purposes only and does not imply an endorsement by Mometrix of any particular political, scientific, or religious point of view.

equation $x + xy - y = 5$ is not a linear equation because it includes the term xy, where two variables are multiplied together.

Linear equations can always be solved (or shown to have no solution) by combining like terms and performing simple operations on both sides of the equation. Some non-linear equations can be solved by similar methods, but others may require more advanced methods of solution, if they can be solved analytically at all.

Solving Equations Involving Roots

In an equation involving roots, the first step is to isolate the term with the root, if possible, and then raise both sides of the equation to the appropriate power to eliminate it. Consider an example equation, $2\sqrt{x+1} - 1 = 3$. In this case, begin by adding 1 to both sides, yielding $2\sqrt{x+1} = 4$, and then dividing both sides by 2, yielding $\sqrt{x+1} = 2$. Now square both sides, yielding $x + 1 = 4$. Finally, subtracting 1 from both sides yields $x = 3$.

Squaring both sides of an equation may, however, yield a spurious solution—a solution to the squared equation that is *not* a solution of the original equation. It's therefore necessary to plug the solution back into the original equation to make sure it works. In this case, it does: $2\sqrt{3+1} - 1 = 2\sqrt{4} - 1 = 2(2) - 1 = 4 - 1 = 3$.

The same procedure applies for other roots as well. For example, given the equation $3 + \sqrt[3]{2x} = 5$, we can first subtract 3 from both sides, yielding $\sqrt[3]{2x} = 2$ and isolating the root. Raising both sides to the third power yields $2x = 2^3$; i.e., $2x = 8$. We can now divide both sides by 2 to get $x = 4$.

Review Video: Solving Equations Involving Roots
Visit mometrix.com/academy and enter code: 297670

Solving Equations with Exponents

To solve an equation involving an exponent, the first step is to isolate the variable with the exponent. We can then take the appropriate root of both sides to eliminate the exponent. For instance, for the equation $2x^3 + 17 = 5x^3 - 7$, we can subtract $5x^3$ from both sides to get $-3x^3 + 17 = -7$, and then subtract 17 from both sides to get $-3x^3 = -24$. Finally, we can divide both sides by -3 to get $x^3 = 8$. Finally, we can take the cube root of both sides to get $x = \sqrt[3]{8} = 2$.

One important but often overlooked point is that equations with an exponent greater than 1 may have more than one answer. The solution to $x^2 = 9$ isn't simply $x = 3$; it's $x = \pm 3$ (that is, $x = 3$ or $x = -3$). For a slightly more complicated example, consider the equation $(x - 1)^2 - 1 = 3$. Adding 1 to both sides yields $(x - 1)^2 = 4$; taking the square root of both sides yields $x - 1 = 2$. We can then add 1 to both sides to get $x = 3$. However, there's a second solution. We also have the possibility that $x - 1 = -2$, in which case $x = -1$. Both $x = 3$ and $x = -1$ are valid solutions, as can be verified by substituting them both into the original equation.

Review Video: Solving Equations with Exponents
Visit mometrix.com/academy and enter code: 514557

Solving Equations with Absolute Values

When solving an equation with an absolute value, the first step is to isolate the absolute value term. We then consider two possibilities: when the expression inside the absolute value is positive or when it is negative. In the former case, the expression in the absolute value equals the expression on the other side of the equation; in the latter, it equals the additive inverse of that expression—the

Copyright © Mometrix Media. You have been licensed one copy of this document for personal use only. Any other reproduction or redistribution is strictly prohibited. All rights reserved.
This content is provided for test preparation purposes only and does not imply an endorsement by Mometrix of any particular political, scientific, or religious point of view.

expression times negative one. We consider each case separately and finally check for spurious solutions.

For instance, consider solving $|2x - 1| + x = 5$ for x. We can first isolate the absolute value by moving the x to the other side: $|2x - 1| = -x + 5$. Now, we have two possibilities. First, that $2x - 1$ is positive, and hence $2x - 1 = -x + 5$. Rearranging and combining like terms yields $3x = 6$, and hence $x = 2$. The other possibility is that $2x - 1$ is negative, and hence $2x - 1 = -(-x + 5) = x - 5$. In this case, rearranging and combining like terms yields $x = -4$. Substituting $x = 2$ and $x = -4$ back into the original equation, we see that they are both valid solutions.

Note that the absolute value of a sum or difference applies to the sum or difference as a whole, not to the individual terms; in general, $|2x - 1|$ is not equal to $|2x + 1|$ or to $|2x| - 1$.

Spurious Solutions

A **spurious solution** may arise when we square both sides of an equation as a step in solving it or under certain other operations on the equation. It is a solution to the squared or otherwise modified equation that is *not* a solution of the original equation. To identify a spurious solution, it's useful when you solve an equation involving roots or absolute values to plug the solution back into the original equation to make sure it's valid.

Choosing Which Variable to Isolate in Two-Variable Equations

Similar to methods for a one-variable equation, solving a two-variable equation involves isolating a variable: manipulating the equation so that a variable appears by itself on one side of the equation, and not at all on the other side. However, in a two-variable equation, you will usually only be able to isolate one of the variables; the other variable may appear on the other side along with constant terms, or with exponents or other functions.

Often one variable will be much more easily isolated than the other, and therefore that's the variable you should choose. If one variable appears with various exponents, and the other is only raised to the first power, the latter variable is the one to isolate: given the equation $a^2 + 2b = a^3 + b + 3$, the b only appears to the first power, whereas a appears squared and cubed, so b is the variable that can be solved for: combining like terms and isolating the b on the left side of the equation, we get $b = a^3 - a^2 + 3$. If both variables are equally easy to isolate, then it's best to isolate the dependent variable, if one is defined; if the two variables are x and y, the convention is that y is the dependent variable.

Review Video: Solving Equations with Variables on Both Sides
Visit mometrix.com/academy and enter code: 402497

Cross Multiplication

Finding an Unknown in Equivalent Expressions

It is often necessary to apply information given about a rate or proportion to a new scenario. For example, if you know that Jedha can run a marathon (26.2 miles) in 3 hours, how long would it take her to run 10 miles at the same pace? Start by setting up equivalent expressions:

$$\frac{26.2 \text{ mi}}{3 \text{ hr}} = \frac{10 \text{ mi}}{x \text{ hr}}$$

Copyright © Mometrix Media. You have been licensed one copy of this document for personal use only. Any other reproduction or redistribution is strictly prohibited. All rights reserved.
This content is provided for test preparation purposes only and does not imply an endorsement by Mometrix of any particular political, scientific, or religious point of view.

Now, cross multiply and solve for x:

$$26.2x = 30$$
$$x = \frac{30}{26.2} = \frac{15}{13.1}$$
$$x \approx 1.15 \text{ hrs } or \text{ 1 hr 9 min}$$

So, at this pace, Jedha could run 10 miles in about 1.15 hours or about 1 hour and 9 minutes.

Review Video: Cross Multiplying Fractions
Visit mometrix.com/academy and enter code: 893904

Graphing Equations

Graphical Solutions to Equations

When equations are shown graphically, they are usually shown on a **Cartesian coordinate plane**. The Cartesian coordinate plane consists of two number lines placed perpendicular to each other and intersecting at the zero point, also known as the origin. The horizontal number line is known as the x-axis, with positive values to the right of the origin, and negative values to the left of the origin. The vertical number line is known as the y-axis, with positive values above the origin, and negative values below the origin. Any point on the plane can be identified by an ordered pair in the form (x, y), called coordinates. The x-value of the coordinate is called the abscissa, and the y-value of the coordinate is called the ordinate. The two number lines divide the plane into **four quadrants**: I, II, III, and IV.

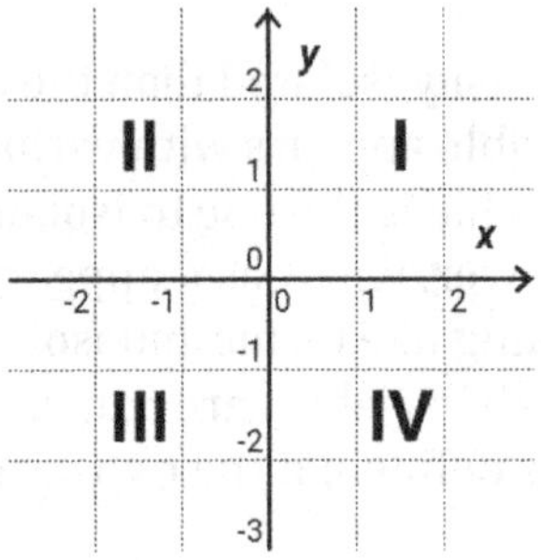

Note that in quadrant I $x > 0$ and $y > 0$, in quadrant II $x < 0$ and $y > 0$, in quadrant III $x < 0$ and $y < 0$, and in quadrant IV $x > 0$ and $y < 0$.

Recall that if the value of the slope of a line is positive, the line slopes upward from left to right. If the value of the slope is negative, the line slopes downward from left to right. If the y-coordinates are the same for two points on a line, the slope is 0 and the line is a **horizontal line**. If the x-coordinates are the same for two points on a line, there is no slope and the line is a **vertical line**. Two or more lines that have equivalent slopes are **parallel lines. Perpendicular lines** have slopes that are negative reciprocals of each other, such as $\frac{a}{b}$ and $\frac{-b}{a}$.

Review Video: Cartesian Coordinate Plane and Graphing
Visit mometrix.com/academy and enter code: 115173

Copyright © Mometrix Media. You have been licensed one copy of this document for personal use only. Any other reproduction or redistribution is strictly prohibited. All rights reserved.
This content is provided for test preparation purposes only and does not imply an endorsement by Mometrix of any particular political, scientific, or religious point of view.

Graphing Equations in Two Variables

One way of graphing an equation in two variables is to plot enough points to get an idea for its shape and then draw the appropriate curve through those points. A point can be plotted by substituting in a value for one variable and solving for the other. If the equation is linear, we only need two points and can then draw a straight line between them.

For example, consider the equation $y = 2x - 1$. This is a linear equation—both variables only appear raised to the first power—so we only need two points. When $x = 0, y = 2(0) - 1 = -1$. When $x = 2, y = 2(2) - 1 = 3$. We can therefore choose the points $(0, -1)$ and $(2, 3)$, and draw a line between them:

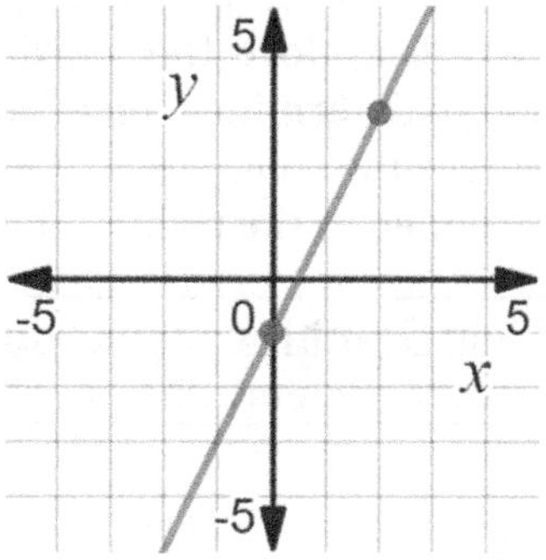

Inequalities

Working with Inequalities

Commonly in algebra and other upper-level fields of math you find yourself working with mathematical expressions that do not equal each other. The statement comparing such expressions with symbols such as $<$ (less than) or $>$ (greater than) is called an *inequality*. An example of an inequality is $7x > 5$. To solve for x, simply divide both sides by 7 and the solution is shown to be $x > \frac{5}{7}$. Graphs of the solution set of inequalities are represented on a number line. Open circles are used to show that an expression approaches a number but is never quite equal to that number.

Review Video: Solving Multi-Step Inequalities
Visit mometrix.com/academy and enter code: 347842

Review Video: Solving Inequalities Using All 4 Basic Operations
Visit mometrix.com/academy and enter code: 401111

Conditional inequalities are those with certain values for the variable that will make the condition true and other values for the variable where the condition will be false. **Absolute inequalities** can have any real number as the value for the variable to make the condition true, while there is no real number value for the variable that will make the condition false. Solving inequalities is done by following the same rules for solving equations with the exception that when multiplying or dividing by a negative number the direction of the inequality sign must be flipped or reversed. **Double**

Copyright © Mometrix Media. You have been licensed one copy of this document for personal use only. Any other reproduction or redistribution is strictly prohibited. All rights reserved.
This content is provided for test preparation purposes only and does not imply an endorsement by Mometrix of any particular political, scientific, or religious point of view.

inequalities are situations where two inequality statements apply to the same variable expression. Example: $-c < ax + b < c$.

Review Video: Conditional and Absolute Inequalities
Visit mometrix.com/academy and enter code: 980164

Determining Solutions to Inequalities

To determine whether a coordinate is a solution of an inequality, you can substitute the values of the coordinate into the inequality, simplify, and check whether the resulting statement holds true. For instance, to determine whether $(-2,4)$ is a solution of the inequality $y \geq -2x + 3$, substitute the values into the inequality, $4 \geq -2(-2) + 3$. Simplify the right side of the inequality and the result is $4 \geq 7$, which is a false statement. Therefore, the coordinate is not a solution of the inequality. You can also use this method to determine which part of the graph of an inequality is shaded. The graph of $y \geq -2x + 3$ includes the solid line $y = -2x + 3$ and, since it excludes the point $(-2,4)$ to the left of the line, it is shaded to the right of the line.

Review Video: Graphing Linear Inequalities
Visit mometrix.com/academy and enter code: 439421

Flipping Inequality Signs

When given an inequality, we can always turn the entire inequality around, swapping the two sides of the inequality and changing the inequality sign. For instance, $x + 2 > 2x - 3$ is equivalent to $2x - 3 < x + 2$. Aside from that, normally the inequality does not change if we carry out the same operation on both sides of the inequality. There is, however, one principal exception: if we *multiply* or *divide* both sides of the inequality by a *negative number*, the inequality is flipped. For example, if we take the inequality $-2x < 6$ and divide both sides by -2, the inequality flips and we are left with $x > -3$. This *only* applies to multiplication and division, and only with negative numbers. Multiplying or dividing both sides by a positive number, or adding or subtracting any number regardless of sign, does not flip the inequality. Another special case that flips the inequality sign is when reciprocals are used. For instance, $3 > 2$ but the relation of the reciprocals is $\frac{1}{2} < \frac{1}{3}$.

Compound Inequalities

A **compound inequality** is an equality that consists of two inequalities combined with *and* or *or*. The two components of a proper compound inequality must be of opposite type: that is, one must be greater than (or greater than or equal to), the other less than (or less than or equal to). For instance, "$x + 1 < 2$ or $x + 1 > 3$" is a compound inequality, as is "$2x \geq 4$ and $2x \leq 6$." An *and* inequality can be written more compactly by having one inequality on each side of the common part: "$2x \geq 1$ and $2x \leq 6$," can also be written as $1 \leq 2x \leq 6$.

In order for the compound inequality to be meaningful, the two parts of an *and* inequality must overlap; otherwise, no numbers satisfy the inequality. On the other hand, if the two parts of an *or* inequality overlap, then *all* numbers satisfy the inequality and as such the inequality is usually not meaningful.

Solving a compound inequality requires solving each part separately. For example, given the compound inequality "$x + 1 < 2$ or $x + 1 > 3$," the first inequality, $x + 1 < 2$, reduces to $x < 1$, and

Copyright © Mometrix Media. You have been licensed one copy of this document for personal use only. Any other reproduction or redistribution is strictly prohibited. All rights reserved.
This content is provided for test preparation purposes only and does not imply an endorsement by Mometrix of any particular political, scientific, or religious point of view.

the second part, $x + 1 > 3$, reduces to $x > 2$, so the whole compound inequality can be written as "$x < 1$ or $x > 2$." Similarly, $1 \leq 2x \leq 6$ can be solved by dividing each term by 2, yielding $\frac{1}{2} \leq x \leq 3$.

Review Video: Compound Inequalities
Visit mometrix.com/academy and enter code: 786318

Solving Inequalities Involving Absolute Values

To solve an inequality involving an absolute value, first isolate the term with the absolute value. Then proceed to treat the two cases separately as with an absolute value equation, but flipping the inequality in the case where the expression in the absolute value is negative (since that essentially involves multiplying both sides by −1.) The two cases are then combined into a compound inequality; if the absolute value is on the greater side of the inequality, then it is an *or* compound inequality, if on the lesser side, then it's an *and*.

Consider the inequality $2 + |x - 1| \geq 3$. We can isolate the absolute value term by subtracting 2 from both sides: $|x - 1| \geq 1$. Now, we're left with the two cases $x - 1 \geq 1$ or $x - 1 \leq -1$: note that in the latter, negative case, the inequality is flipped. $x - 1 \geq 1$ reduces to $x \geq 2$, and $x - 1 \leq -1$ reduces to $x \leq 0$. Since in the inequality $|x - 1| \geq 1$ the absolute value is on the greater side, the two cases combine into an *or* compound inequality, so the final, solved inequality is "$x \leq 0$ or $x \geq 2$."

Review Video: Solving Absolute Value Inequalities
Visit mometrix.com/academy and enter code: 997008

Solving Inequalities Involving Square Roots

Solving an inequality with a square root involves two parts. First, we solve the inequality as if it were an equation, isolating the square root and then squaring both sides of the equation. Second, we restrict the solution to the set of values of x for which the value inside the square root sign is non-negative.

For example, in the inequality, $\sqrt{x - 2} + 1 < 5$, we can isolate the square root by subtracting 1 from both sides, yielding $\sqrt{x - 2} < 4$. Squaring both sides of the inequality yields $x - 2 < 16$, so $x < 18$. Since we can't take the square root of a negative number, we also require the part inside the square root to be non-negative. In this case, that means $x - 2 \geq 0$. Adding 2 to both sides of the inequality yields $x \geq 2$. Our final answer is a compound inequality combining the two simple inequalities: $x \geq 2$ and $x < 18$, or $2 \leq x < 18$.

Note that we only get a compound inequality if the two simple inequalities are in opposite directions; otherwise, we take the one that is more restrictive.

The same technique can be used for other even roots, such as fourth roots. It is *not*, however, used for cube roots or other odd roots—negative numbers *do* have cube roots, so the condition that the quantity inside the root sign cannot be negative does not apply.

Review Video: Solving Inequalities Involving Square Roots
Visit mometrix.com/academy and enter code: 800288

Special Circumstances

Sometimes an inequality involving an absolute value or an even exponent is true for all values of x, and we don't need to do any further work to solve it. This is true if the inequality, once the absolute

Copyright © Mometrix Media. You have been licensed one copy of this document for personal use only. Any other reproduction or redistribution is strictly prohibited. All rights reserved.
This content is provided for test preparation purposes only and does not imply an endorsement by Mometrix of any particular political, scientific, or religious point of view.

value or exponent term is isolated, says that term is greater than a negative number (or greater than or equal to zero). Since an absolute value or a number raised to an even exponent is *always* non-negative, this inequality is always true.

Graphical Solutions to Inequalities

Graphing Simple Inequalities

To graph a simple inequality, we first mark on the number line the value that signifies the end point of the inequality. If the inequality is strict (involves a less than or greater than), we use a hollow circle; if it is not strict (less than or equal to or greater than or equal to), we use a solid circle. We then fill in the part of the number line that satisfies the inequality: to the left of the marked point for less than (or less than or equal to), to the right for greater than (or greater than or equal to).

For example, we would graph the inequality $x < 5$ by putting a hollow circle at 5 and filling in the part of the line to the left:

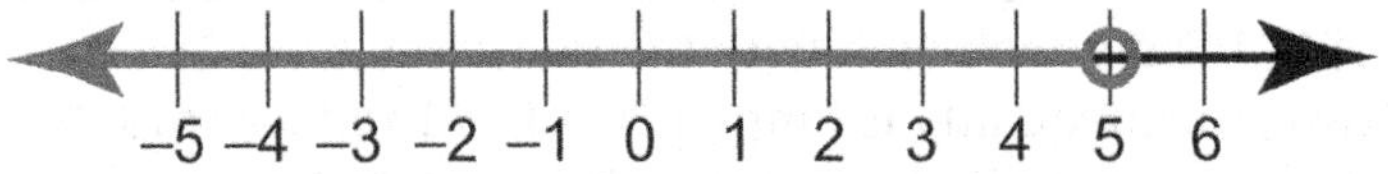

Graphing Compound Inequalities

To graph a compound inequality, we fill in both parts of the inequality for an *or* inequality, or the overlap between them for an *and* inequality. More specifically, we start by plotting the endpoints of each inequality on the number line. For an *or* inequality, we then fill in the appropriate side of the line for each inequality. Typically, the two component inequalities do not overlap, which means the shaded part is *outside* the two points. For an *and* inequality, we instead fill in the part of the line that meets both inequalities.

For the inequality "$x \leq -3$ or $x > 4$," we first put a solid circle at –3 and a hollow circle at 4. We then fill the parts of the line *outside* these circles:

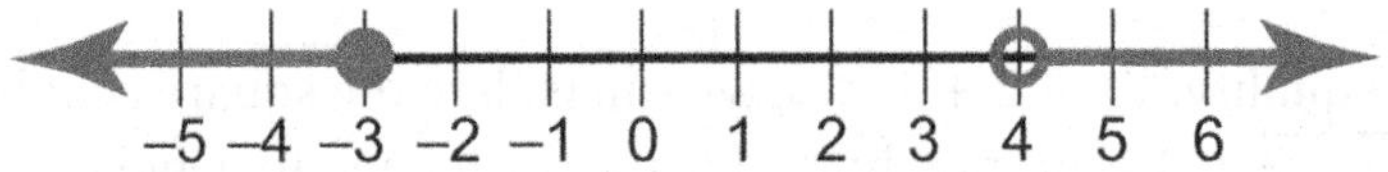

Graphing Inequalities Including Absolute Values

An inequality with an absolute value can be converted to a compound inequality. To graph the inequality, first convert it to a compound inequality, and then graph that normally. If the absolute value is on the greater side of the inequality, we end up with an *or* inequality; we plot the endpoints of the inequality on the number line and fill in the part of the line *outside* those points. If the absolute value is on the smaller side of the inequality, we end up with an *and* inequality; we plot the endpoints of the inequality on the number line and fill in the part of the line *between* those points.

For example, the inequality $|x + 1| \geq 4$ can be rewritten as $x \geq 3$ or $x \leq -5$. We place solid circles at the points 3 and –5 and fill in the part of the line *outside* them:

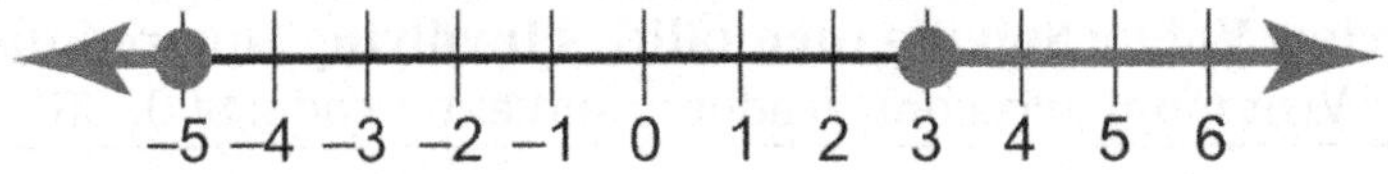

Graphing Inequalities in Two Variables

To graph an inequality in two variables, we first graph the border of the inequality. This means graphing the equation that we get if we replace the inequality sign with an equals sign. If the

Copyright © Mometrix Media. You have been licensed one copy of this document for personal use only. Any other reproduction or redistribution is strictly prohibited. All rights reserved.
This content is provided for test preparation purposes only and does not imply an endorsement by Mometrix of any particular political, scientific, or religious point of view.

inequality is strict (> or <), we graph the border with a dashed or dotted line; if it is not strict (≥ or ≤), we use a solid line. We can then test any point not on the border to see if it satisfies the inequality. If it does, we shade in that side of the border; if not, we shade in the other side. As an example, consider $y > 2x + 2$. To graph this inequality, we first graph the border, $y = 2x + 2$. Since it is a strict inequality, we use a dashed line. Then, we choose a test point. This can be any point not on the border; in this case, we will choose the origin, (0,0). (This makes the calculation easy and is generally a good choice unless the border passes through the origin.) Putting this into the original inequality, we get $0 > 2(0) + 2$, i.e., $0 > 2$. This is *not* true, so we shade in the side of the border that does *not* include the point (0,0):

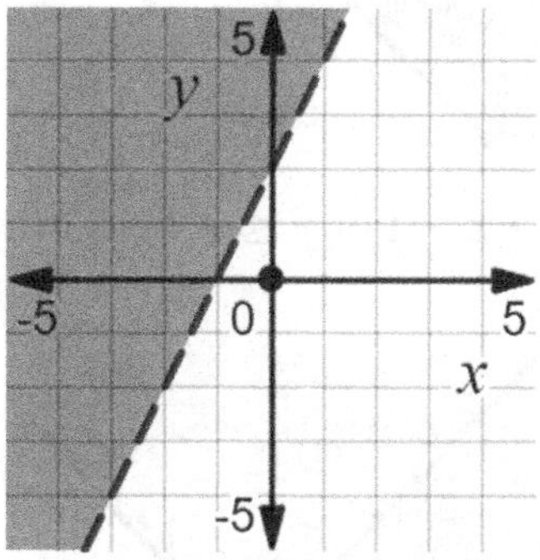

Graphing Compound Inequalities in Two Variables

One way to graph a compound inequality in two variables is to first graph each of the component inequalities. For an *and* inequality, we then shade in only the parts where the two graphs overlap; for an *or* inequality, we shade in any region that pertains to either of the individual inequalities.

Consider the graph of "$y \geq x - 1$ and $y \leq -x$":

We first shade in the individual inequalities:

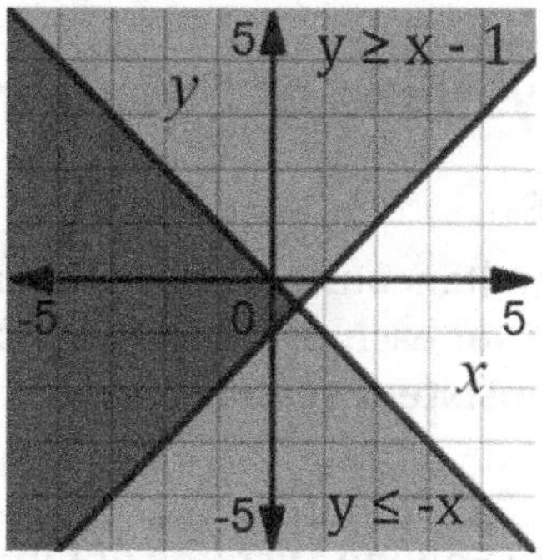

Copyright © Mometrix Media. You have been licensed one copy of this document for personal use only. Any other reproduction or redistribution is strictly prohibited. All rights reserved.
This content is provided for test preparation purposes only and does not imply an endorsement by Mometrix of any particular political, scientific, or religious point of view.

Now, since the compound inequality has an *and*, we only leave shaded the overlap—the part that pertains to *both* inequalities:

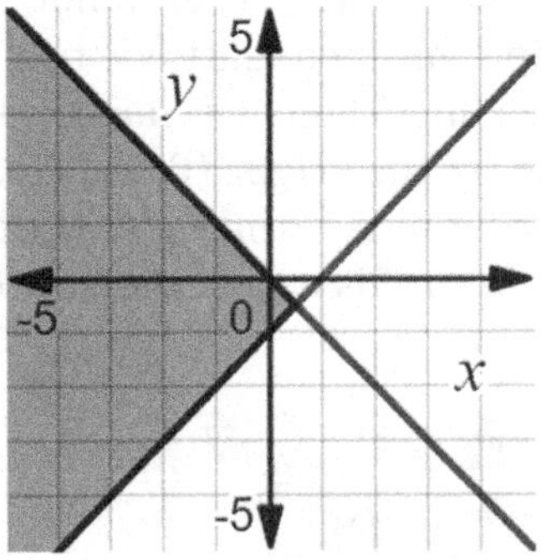

If instead the inequality had been "$y \geq x - 1$ or $y \leq -x$," our final graph would involve the *total* shaded area:

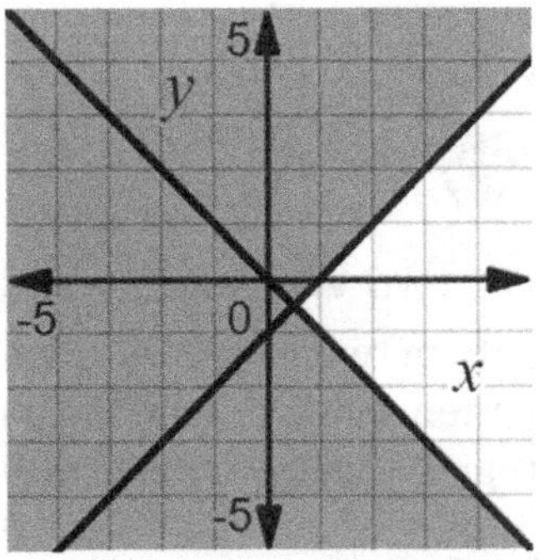

Review Video: Graphing Solutions to Inequalities
Visit mometrix.com/academy and enter code: 391281

Systems of Equations

Solving Systems of Equations

A **system of equations** is a set of simultaneous equations that all use the same variables. A solution to a system of equations must be true for each equation in the system. **Consistent systems** are those with at least one solution. **Inconsistent systems** are systems of equations that have no solution.

Review Video: Solving Systems of Linear Equations
Visit mometrix.com/academy and enter code: 746745

Substitution

To solve a system of linear equations by **substitution**, start with the easier equation and solve for one of the variables. Express this variable in terms of the other variable. Substitute this expression in the other equation and solve for the other variable. The solution should be expressed in the form (x, y). Substitute the values into both of the original equations to check your answer. Consider the following system of equations:

$$x + 6y = 15$$
$$3x - 12y = 18$$

Copyright © Mometrix Media. You have been licensed one copy of this document for personal use only. Any other reproduction or redistribution is strictly prohibited. All rights reserved.
This content is provided for test preparation purposes only and does not imply an endorsement by Mometrix of any particular political, scientific, or religious point of view.

Solving the first equation for x: $x = 15 - 6y$

Substitute this value in place of x in the second equation, and solve for y:

$$\begin{aligned} 3(15 - 6y) - 12y &= 18 \\ 45 - 18y - 12y &= 18 \\ 30y &= 27 \\ y &= \frac{27}{30} = \frac{9}{10} = 0.9 \end{aligned}$$

Plug this value for y back into the first equation to solve for x:

$$x = 15 - 6(0.9) = 15 - 5.4 = 9.6$$

Check both equations if you have time:

$$\begin{aligned} 9.6 + 6(0.9) &= 15 \\ 9.6 + 5.4 &= 15 \\ 15 &= 15 \end{aligned} \qquad \begin{aligned} 3(9.6) - 12(0.9) &= 18 \\ 28.8 - 10.8 &= 18 \\ 18 &= 18 \end{aligned}$$

Therefore, the solution is $(9.6, 0.9)$.

Review Video: The Substitution Method
Visit mometrix.com/academy and enter code: 565151

Review Video: Substitution and Elimination
Visit mometrix.com/academy and enter code: 958611

ELIMINATION

To solve a system of equations using **elimination**, begin by rewriting both equations in standard form $Ax + By = C$. Check to see if the coefficients of one pair of like variables add to zero. If not, multiply one or both of the equations by a non-zero number to make one set of like variables add to zero. Add the two equations to solve for one of the variables. Substitute this value into one of the original equations to solve for the other variable. Check your work by substituting into the other equation. Now, let's look at solving the following system using the elimination method:

$$\begin{aligned} 5x + 6y &= 4 \\ x + 2y &= 4 \end{aligned}$$

If we multiply the second equation by -3, we can eliminate the y-terms:

$$\begin{aligned} 5x + 6y &= 4 \\ -3x - 6y &= -12 \end{aligned}$$

Add the equations together and solve for x:

$$\begin{aligned} 2x &= -8 \\ x &= \frac{-8}{2} = -4 \end{aligned}$$

Copyright © Mometrix Media. You have been licensed one copy of this document for personal use only. Any other reproduction or redistribution is strictly prohibited. All rights reserved.
This content is provided for test preparation purposes only and does not imply an endorsement by Mometrix of any particular political, scientific, or religious point of view.

Plug the value for x back in to either of the original equations and solve for y:

$$-4 + 2y = 4$$
$$y = \frac{4 + 4}{2} = 4$$

Check both equations if you have time:

$$5(-4) + 6(4) = 4 \qquad -4 + 2(4) = 4$$
$$-20 + 24 = 4 \qquad -4 + 8 = 4$$
$$4 = 4 \qquad 4 = 4$$

Therefore, the solution is $(-4,4)$.

Review Video: The Elimination Method
Visit mometrix.com/academy and enter code: 449121

GRAPHICALLY

To solve a system of linear equations **graphically**, plot both equations on the same graph. The solution of the equations is the point where both lines cross. If the lines do not cross (are parallel), then there is **no solution**.

For example, consider the following system of equations:

$$y = 2x + 7$$
$$y = -x + 1$$

Since these equations are given in slope-intercept form, they are easy to graph; the y-intercepts of the lines are $(0,7)$ and $(0,1)$. The respective slopes are 2 and -1, thus the graphs look like this:

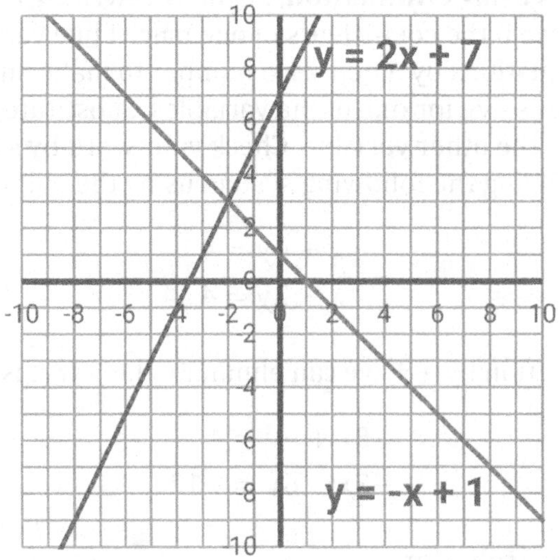

The two lines intersect at the point $(-2,3)$, thus this is the solution to the system of equations.

Solving a system graphically is generally only practical if both coordinates of the solution are integers; otherwise the intersection will lie between gridlines on the graph and the coordinates will be difficult or impossible to determine exactly. It also helps if, as in this example, the equations are

Copyright © Mometrix Media. You have been licensed one copy of this document for personal use only. Any other reproduction or redistribution is strictly prohibited. All rights reserved.
This content is provided for test preparation purposes only and does not imply an endorsement by Mometrix of any particular political, scientific, or religious point of view.

in slope-intercept form or some other form that makes them easy to graph. Otherwise, another method of solution (by substitution or elimination) is likely to be more useful.

Review Video: Solving Systems by Graphing
Visit mometrix.com/academy and enter code: 634812

Solving Systems of Equations Using the Trace Feature

Using the trace feature on a calculator requires that you rewrite each equation, isolating the y-variable on one side of the equal sign. Enter both equations in the graphing calculator and plot the graphs simultaneously. Use the trace cursor to find where the two lines cross. Use the zoom feature if necessary to obtain more accurate results. Always check your answer by substituting into the original equations. The trace method is likely to be less accurate than other methods due to the resolution of graphing calculators but is a useful tool to provide an approximate answer.

Advanced Systems of Equations

Solving a System of Equations Consisting of a Linear Equation and a Quadratic Equation

Algebraically

Generally, the simplest way to solve a system of equations consisting of a linear equation and a quadratic equation algebraically is through the method of substitution. One possible strategy is to solve the linear equation for y and then substitute that expression into the quadratic equation. After expansion and combining like terms, this will result in a new quadratic equation for x, which, like all quadratic equations, may have zero, one, or two solutions. Plugging each solution for x back into one of the original equations will then produce the corresponding value of y.

For example, consider the following system of equations:

$$x + y = 1$$
$$y = (x + 3)^2 - 2$$

We can solve the linear equation for y to yield $y = -x + 1$. Substituting this expression into the quadratic equation produces $-x + 1 = (x + 3)^2 - 2$. We can simplify this equation:

$$\begin{aligned} -x + 1 &= (x + 3)^2 - 2 \\ -x + 1 &= x^2 + 6x + 9 - 2 \\ -x + 1 &= x^2 + 6x + 7 \\ 0 &= x^2 + 7x + 6 \end{aligned}$$

This quadratic equation can be factored as $(x + 1)(x + 6) = 0$. It therefore has two solutions: $x_1 = -1$ and $x_2 = -6$. Plugging each of these back into the original linear equation yields $y_1 = -x_1 + 1 = -(-1) + 1 = 2$ and $y_2 = -x_2 + 1 = -(-6) + 1 = 7$. Thus, this system of equations has two solutions, $(-1,2)$ and $(-6,7)$.

It may help to check your work by putting each x- and y-value back into the original equations and verifying that they do provide a solution.

Graphically

To solve a system of equations consisting of a linear equation and a quadratic equation graphically, plot both equations on the same graph. The linear equation will, of course, produce a straight line,

Copyright © Mometrix Media. You have been licensed one copy of this document for personal use only. Any other reproduction or redistribution is strictly prohibited. All rights reserved.
This content is provided for test preparation purposes only and does not imply an endorsement by Mometrix of any particular political, scientific, or religious point of view.

while the quadratic equation will produce a parabola. These two graphs will intersect at zero, one, or two points; each point of intersection is a solution of the system.

For example, consider the following system of equations:

$$y = -2x + 2$$
$$y = -2x^2 + 4x + 2$$

The linear equation describes a line with a y-intercept of $(0,2)$ and a slope of -2.

To graph the quadratic equation, we can first find the vertex of the parabola: the x-coordinate of the vertex is $h = -\frac{b}{2a} = -\frac{4}{2(-2)} = 1$, and the y-coordinate is $k = -2(1)^2 + 4(1) + 2 = 4$. Thus, the vertex lies at $(1,4)$. To get a feel for the rest of the parabola, we can plug in a few more values of x to find more points; by putting in $x = 2$ and $x = 3$ in the quadratic equation, we find that the points $(2,2)$ and $(3,-4)$ lie on the parabola; by symmetry, so must $(0,2)$ and $(-1,-4)$. We can now plot both equations:

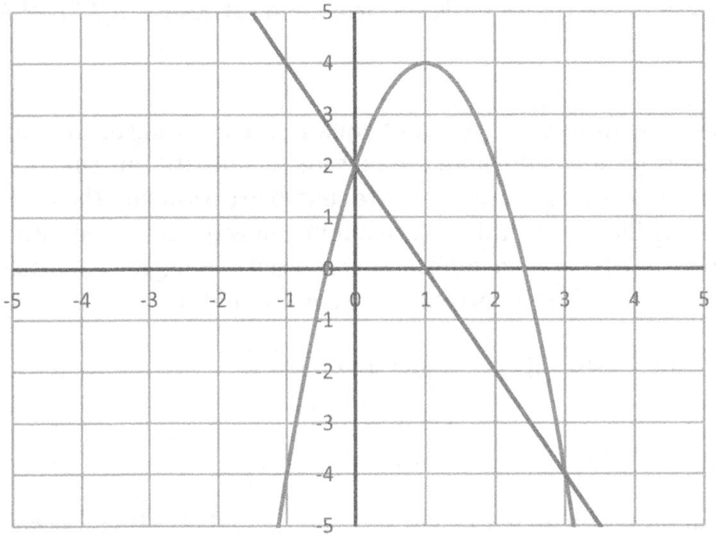

These two curves intersect at the points $(0,2)$ and $(3,-4)$, thus these are the solutions of the equation.

Review Video: Solving a System of Linear and Quadratic Equations
Visit mometrix.com/academy and enter code: 194870

Calculations Using Points

Sometimes you need to perform calculations using only points on a graph as input data. Using points, you can determine what the **midpoint** and **distance** are. If you know the equation for a line, you can calculate the distance between the line and the point.

To find the **midpoint** of two points (x_1, y_1) and (x_2, y_2), average the x-coordinates to get the x-coordinate of the midpoint, and average the y-coordinates to get the y-coordinate of the midpoint. The formula is: $\left(\frac{x_1+x_2}{2}, \frac{y_1+y_2}{2}\right)$.

Copyright © Mometrix Media. You have been licensed one copy of this document for personal use only. Any other reproduction or redistribution is strictly prohibited. All rights reserved.
This content is provided for test preparation purposes only and does not imply an endorsement by Mometrix of any particular political, scientific, or religious point of view.

The **distance** between two points is the same as the length of the hypotenuse of a right triangle with the two given points as endpoints, and the two sides of the right triangle parallel to the x-axis and y-axis, respectively. The length of the segment parallel to the x-axis is the difference between the x-coordinates of the two points. The length of the segment parallel to the y-axis is the difference between the y-coordinates of the two points. Use the Pythagorean theorem $a^2 + b^2 = c^2$ or $c = \sqrt{a^2 + b^2}$ to find the distance. The formula is $d = \sqrt{(x_2 - x_1)^2 + (y_2 - y_1)^2}$.

When a line is in the format $Ax + By + C = 0$, where A, B, and C are coefficients, you can use a point (x_1, y_1) not on the line and apply the formula $d = \frac{|Ax_1 + By_1 + C|}{\sqrt{A^2 + B^2}}$ to find the distance between the line and the point (x_1, y_1).

Review Video: Calculations Using Points on a Graph
Visit mometrix.com/academy and enter code: 883228

Polynomials

Monomials and Polynomials

A **monomial** is a single constant, variable, or product of constants and variables, such as 7, x, $2x$, or x^3y. There will never be addition or subtraction symbols in a monomial. Like monomials have like variables, but they may have different coefficients. **Polynomials** are algebraic expressions that use addition and subtraction to combine two or more monomials. Two terms make a **binomial**, three terms make a **trinomial**, etc. The **degree of a monomial** is the sum of the exponents of the variables. The **degree of a polynomial** is the highest degree of any individual term.

Review Video: Polynomials
Visit mometrix.com/academy and enter code: 305005

Simplifying Polynomials

Simplifying polynomials requires combining like terms. The like terms in a polynomial expression are those that have the same variable raised to the same power. It is often helpful to connect the like terms with arrows or lines in order to separate them from the other monomials. Once you have determined the like terms, you can rearrange the polynomial by placing them together. Remember to include the sign that is in front of each term. Once the like terms are placed together, you can apply each operation and simplify. When adding and subtracting polynomials, only add and subtract the **coefficient**, or the number part; the variable and exponent stay the same.

Add Polynomials

To add polynomials, you need to add like terms. These terms have the same variable part. An example is $4x^2$ and $3x^2$ have x^2 terms. To find the sum of like terms, find the sum of the coefficients. Then, keep the same variable part. You can use the distributive property to distribute the plus sign to each term of the polynomial. For example:

$(4x^2 - 5x + 7) + (3x^2 + 2x + 1) =$
$(4x^2 - 5x + 7) + 3x^2 + 2x + 1 =$
$(4x^2 + 3x^2) + (-5x + 2x) + (7 + 1) =$
$7x^2 - 3x + 8$

Copyright © Mometrix Media. You have been licensed one copy of this document for personal use only. Any other reproduction or redistribution is strictly prohibited. All rights reserved.
This content is provided for test preparation purposes only and does not imply an endorsement by Mometrix of any particular political, scientific, or religious point of view.

SUBTRACT POLYNOMIALS

To subtract polynomials, you need to subtract like terms. To find the difference of like terms, find the difference of the coefficients. Then, keep the same variable part. You can use the distributive property to distribute the minus sign to each term of the polynomial. For example:

$(-2x^2 - x + 5) - (3x^2 - 4x + 1) =$
$(-2x^2 - x + 5) - 3x^2 + 4x - 1 =$
$(-2x^2 - 3x^2) + (-x + 4x) + (5 - 1) =$
$-5x^2 + 3x + 4$

Review Video: Adding and Subtracting Polynomials
Visit mometrix.com/academy and enter code: 124088

MULTIPLYING POLYNOMIALS

In general, multiplying polynomials is done by multiplying each term in one polynomial by each term in the other and adding the results. In the specific case for multiplying binomials, there is a useful acronym, FOIL, that can help you make sure to cover each combination of terms. The **FOIL method** for $(Ax + By)(Cx + Dy)$ would be:

F	Multiply the *first* terms of each binomial	$(\overbrace{Ax}^{first} + By)(\overbrace{Cx}^{first} + Dy)$	ACx^2
O	Multiply the *outer* terms	$(\overbrace{Ax}^{outer} + By)(Cx + \overbrace{Dy}^{outer})$	$ADxy$
I	Multiply the *inner* terms	$(Ax + \overbrace{By}^{inner})(\overbrace{Cx}^{inner} + Dy)$	$BCxy$
L	Multiply the *last* terms of each binomial	$(Ax + \overbrace{By}^{last})(Cx + \overbrace{Dy}^{last})$	BDy^2

Then, add up the result of each and combine like terms: $ACx^2 + (AD + BC)xy + BDy^2$.

For example, using the FOIL method on binomials $(x + 2)$ and $(x - 3)$:

$$\begin{aligned} &\text{First:} && (\boxed{x} + 2)(\boxed{x} + (-3)) && \rightarrow && (x)(x) = x^2 \\ &\text{Outer:} && (\boxed{x} + 2)(x + \boxed{(-3)}) && \rightarrow && (x)(-3) = -3x \\ &\text{Inner:} && (x + \boxed{2})(\boxed{x} + (-3)) && \rightarrow && (2)(x) = 2x \\ &\text{Last:} && (x + \boxed{2})(x + \boxed{(-3)}) && \rightarrow && (2)(-3) = -6 \end{aligned}$$

This results in: $(x^2) + (-3x) + (2x) + (-6)$

Combine like terms: $x^2 + (-3 + 2)x + (-6) = x^2 - x - 6$

Review Video: Multiplying Terms Using the FOIL Method
Visit mometrix.com/academy and enter code: 854792

DIVIDING POLYNOMIALS

Use long division to divide a polynomial by either a monomial or another polynomial of equal or lesser degree.

When **dividing by a monomial**, divide each term of the polynomial by the monomial.

Copyright © Mometrix Media. You have been licensed one copy of this document for personal use only. Any other reproduction or redistribution is strictly prohibited. All rights reserved.
This content is provided for test preparation purposes only and does not imply an endorsement by Mometrix of any particular political, scientific, or religious point of view.

When **dividing by a polynomial**, begin by arranging the terms of each polynomial in order of one variable. You may arrange in ascending or descending order, but be consistent with both polynomials. To get the first term of the quotient, divide the first term of the dividend by the first term of the divisor. Multiply the first term of the quotient by the entire divisor and subtract that product from the dividend. Repeat for the second and successive terms until you either get a remainder of zero or a remainder whose degree is less than the degree of the divisor. If the quotient has a remainder, write the answer as a mixed expression in the form:

$$\text{quotient} + \frac{\text{remainder}}{\text{divisor}}$$

For example, we can evaluate the following expression in the same way as long division:

$$\frac{x^3 - 3x^2 - 2x + 5}{x - 5}$$

$$
\begin{array}{rl}
 & \quad x^2 \quad + 2x \quad + 8 \\
x - 5 & \overline{)\; x^3 - 3x^2 - 2x + 5} \\
 & \underline{-(x^3 - 5x^2)} \\
 & \qquad 2x^2 - 2x \\
 & \qquad \underline{-(2x^2 - 10x)} \\
 & \qquad\qquad 8x + 5 \\
 & \qquad\qquad \underline{-(8x - 40)} \\
 & \qquad\qquad\quad 45
\end{array}
$$

$$\frac{x^3 - 3x^2 - 2x + 5}{x - 5} = x^2 + 2x + 8 + \frac{45}{x - 5}$$

When **factoring** a polynomial, first check for a common monomial factor, that is, look to see if each coefficient has a common factor or if each term has an x in it. If the factor is a trinomial but not a perfect trinomial square, look for a factorable form, such as one of these:

$$x^2 + (a + b)x + ab = (x + a)(x + b)$$
$$(ac)x^2 + (ad + bc)x + bd = (ax + b)(cx + d)$$

For factors with four terms, look for groups to factor. Once you have found the factors, write the original polynomial as the product of all the factors. Make sure all of the polynomial factors are prime. Monomial factors may be *prime* or *composite*. Check your work by multiplying the factors to make sure you get the original polynomial.

Below are patterns of some special products to remember to help make factoring easier:

- Perfect trinomial squares: $x^2 + 2xy + y^2 = (x + y)^2$ or $x^2 - 2xy + y^2 = (x - y)^2$
- Difference between two squares: $x^2 - y^2 = (x + y)(x - y)$
- Sum of two cubes: $x^3 + y^3 = (x + y)(x^2 - xy + y^2)$
    - Note: the second factor is *not* the same as a perfect trinomial square, so do not try to factor it further.
- Difference between two cubes: $x^3 - y^3 = (x - y)(x^2 + xy + y^2)$

Copyright © Mometrix Media. You have been licensed one copy of this document for personal use only. Any other reproduction or redistribution is strictly prohibited. All rights reserved.
This content is provided for test preparation purposes only and does not imply an endorsement by Mometrix of any particular political, scientific, or religious point of view.

- Again, the second factor is *not* the same as a perfect trinomial square.
- Perfect cubes: $x^3 + 3x^2y + 3xy^2 + y^3 = (x + y)^3$ and $x^3 - 3x^2y + 3xy^2 - y^3 = (x - y)^3$

Rational Expressions

Rational expressions are fractions with polynomials in both the numerator and the denominator; the value of the polynomial in the denominator cannot be equal to zero. Be sure to keep track of values that make the denominator of the original expression zero as the final result inherits the same restrictions. For example, a denominator of $x - 3$ indicates that the expression is not defined when $x = 3$ and, as such, regardless of any operations done to the expression, it remains undefined there.

To **add or subtract** rational expressions, first find the common denominator, then rewrite each fraction as an equivalent fraction with the common denominator. Finally, add or subtract the numerators to get the numerator of the answer, and keep the common denominator as the denominator of the answer.

When **multiplying** rational expressions, factor each polynomial and cancel like factors (a factor which appears in both the numerator and the denominator). Then, multiply all remaining factors in the numerator to get the numerator of the product, and multiply the remaining factors in the denominator to get the denominator of the product. Remember: cancel entire factors, not individual terms.

To **divide** rational expressions, take the reciprocal of the divisor (the rational expression you are dividing by) and multiply by the dividend.

Review Video: Rational Expressions
Visit mometrix.com/academy and enter code: 415183

Simplifying Rational Expressions

To simplify a rational expression, factor the numerator and denominator completely. Factors that are the same and appear in the numerator and denominator have a ratio of 1. For example, look at the following expression:

$$\frac{x - 1}{1 - x^2}$$

The denominator, $(1 - x^2)$, is a difference of squares. It can be factored as $(1 - x)(1 + x)$. The factor $1 - x$ and the numerator $x - 1$ are opposites and have a ratio of –1. Rewrite the numerator as $-1(1 - x)$. So, the rational expression can be simplified as follows:

$$\frac{x - 1}{1 - x^2} = \frac{-1(1 - x)}{(1 - x)(1 + x)} = \frac{-1}{1 + x}$$

Note that since the original expression is only defined for $x \neq \{-1, 1\}$, the simplified expression has the same restrictions.

Review Video: Reducing Rational Expressions
Visit mometrix.com/academy and enter code: 788868

Copyright © Mometrix Media. You have been licensed one copy of this document for personal use only. Any other reproduction or redistribution is strictly prohibited. All rights reserved.
This content is provided for test preparation purposes only and does not imply an endorsement by Mometrix of any particular political, scientific, or religious point of view.

Algebraic Theorems

According to the **fundamental theorem of algebra**, every non-constant, single-variable polynomial has exactly as many roots as the polynomial's highest exponent. For example, if x^4 is the largest exponent of a term, the polynomial will have exactly 4 roots. However, some of these roots may have multiplicity or be complex numbers. For instance, in the polynomial function $f(x) = x^4 - 4x + 3$, the only real root is 1, though it has multiplicity of 2 – that is, it occurs twice. The other two roots, $(-1 - i\sqrt{2})$ and $(-1 + i\sqrt{2})$, are complex, consisting of both real and non-real components.

The **remainder theorem** is useful for determining the remainder when a polynomial is divided by a binomial. The remainder theorem states that if a polynomial function $f(x)$ is divided by a binomial $x - a$, where a is a real number, the remainder of the division will be the value of $f(a)$. If $f(a) = 0$, then a is a root of the polynomial.

The **factor theorem** is related to the remainder theorem and states that if $f(a) = 0$ then $(x - a)$ is a factor of the function.

According to the **rational root theorem,** any rational root of a polynomial function $f(x) = a_n x^n + a_{n-1}x^{n-1} + \cdots + a_1 x + a_0$ with integer coefficients will, when reduced to its lowest terms, be a positive or negative fraction such that the numerator is a factor of a_0 and the denominator is a factor of a_n. For instance, if the polynomial function $f(x) = x^3 + 3x^2 - 4$ has any rational roots, the numerators of those roots can only be factors of 4 (1, 2, 4), and the denominators can only be factors of 1 (1). The function in this example has roots of 1 (or $\frac{1}{1}$) and -2 (or $\frac{-2}{1}$).

Quadratics

Solving Quadratic Equations

Quadratic equations are a special set of trinomials of the form $y = ax^2 + bx + c$ that occur commonly in math and real-world applications. The **roots** of a quadratic equation are the solutions that satisfy the equation when $y = 0$; in other words, where the graph touches the x-axis. There are several ways to determine these solutions including using the quadratic formula, factoring, completing the square, and graphing the function.

Review Video: Quadratic Equations Overview
Visit mometrix.com/academy and enter code: 476276

Review Video: Solutions of a Quadratic Equation on a Graph
Visit mometrix.com/academy and enter code: 328231

Quadratic Formula

The **quadratic formula** is used to solve quadratic equations when other methods are more difficult. To use the quadratic formula to solve a quadratic equation, begin by rewriting the equation in standard form $ax^2 + bx + c = 0$, where a, b, and c are coefficients. Once you have identified the values of the coefficients, substitute those values into the quadratic formula

$$x = \frac{-b \pm \sqrt{b^2 - 4ac}}{2a}$$

Copyright © Mometrix Media. You have been licensed one copy of this document for personal use only. Any other reproduction or redistribution is strictly prohibited. All rights reserved.
This content is provided for test preparation purposes only and does not imply an endorsement by Mometrix of any particular political, scientific, or religious point of view.

Evaluate the equation and simplify the expression. Again, check each root by substituting into the original equation. In the quadratic formula, the portion of the formula under the radical $(b^2 - 4ac)$ is called the **discriminant**. If the discriminant is zero, there is only one root: $-\frac{b}{2a}$. If the discriminant is positive, there are two different real roots. If the discriminant is negative, there are no real roots; you will instead find complex roots. Often these solutions don't make sense in context and are ignored.

Review Video: Using the Quadratic Formula
Visit mometrix.com/academy and enter code: 163102

Factoring

To solve a quadratic equation by factoring, begin by rewriting the equation in standard form, $x^2 + bx + c = 0$. Remember that the goal of factoring is to find numbers f and g such that $(x + f)(x + g) = x^2 + (f + g)x + fg$, in other words $(f + g) = b$ and $fg = c$. This can be a really useful method when b and c are integers. Determine the factors of c and look for pairs that could sum to b.

For example, consider finding the roots of $x^2 + 6x - 16 = 0$. The factors of -16 include, -4 and 4, -8 and 2, -2 and 8, -1 and 16, and 1 and -16. The factors that sum to 6 are -2 and 8. Write these factors as the product of two binomials, $0 = (x - 2)(x + 8)$. Finally, since these binomials multiply together to equal zero, set them each equal to zero and solve each for x. This results in $x - 2 = 0$, which simplifies to $x = 2$ and $x + 8 = 0$, which simplifies to $x = -8$. Therefore, the roots of the equation are 2 and -8.

Review Video: Factoring Quadratic Equations
Visit mometrix.com/academy and enter code: 336566

Completing the Square

One way to find the roots of a quadratic equation is to find a way to manipulate it such that it follows the form of a perfect square $(x^2 + 2px + p^2)$ by adding and subtracting a constant. This process is called **completing the square**. In other words, if you are given a quadratic that is not a perfect square, $x^2 + bx + c = 0$, you can find a constant d that could be added in to make it a perfect square:

$$x^2 + bx + c + (d - d) = 0; \ \{\text{Let } b = 2p \text{ and } c + d = p^2\}$$

then:

$$x^2 + 2px + p^2 - d = 0 \text{ and } d = \frac{b^2}{4} - c$$

Once you have completed the square you can find the roots of the resulting equation:

$$\begin{aligned} x^2 + 2px + p^2 - d &= 0 \\ (x + p)^2 &= d \\ x + p &= \pm\sqrt{d} \\ x &= -p \pm \sqrt{d} \end{aligned}$$

Copyright © Mometrix Media. You have been licensed one copy of this document for personal use only. Any other reproduction or redistribution is strictly prohibited. All rights reserved.
This content is provided for test preparation purposes only and does not imply an endorsement by Mometrix of any particular political, scientific, or religious point of view.

It is worth noting that substituting the original expressions into this solution gives the same result as the quadratic formula where $a = 1$:

$$x = -p \pm \sqrt{d} = -\frac{b}{2} \pm \sqrt{\frac{b^2}{4} - c} = -\frac{b}{2} \pm \frac{\sqrt{b^2 - 4c}}{2} = \frac{-b \pm \sqrt{b^2 - 4c}}{2}$$

Completing the square can be seen as arranging block representations of each of the terms to be as close to a square as possible and then filling in the gaps. For example, consider the quadratic expression $x^2 + 6x + 2$:

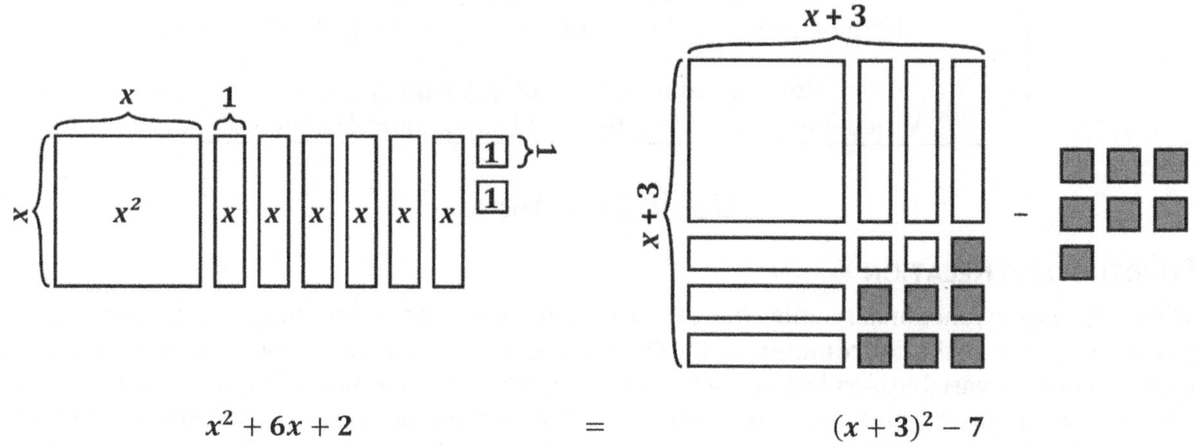

Review Video: Completing the Square
Visit mometrix.com/academy and enter code: 982479

USING GIVEN ROOTS TO FIND QUADRATIC EQUATION

One way to find the roots of a quadratic equation is to factor the equation and use the **zero product property**, setting each factor of the equation equal to zero to find the corresponding root. We can use this technique in reverse to find an equation given its roots. Each root corresponds to a linear equation which in turn corresponds to a factor of the quadratic equation.

For example, we can find a quadratic equation whose roots are $x = 2$ and $x = -1$. The root $x = 2$ corresponds to the equation $x - 2 = 0$, and the root $x = -1$ corresponds to the equation $x + 1 = 0$.

These two equations correspond to the factors $(x - 2)$ and $(x + 1)$, from which we can derive the equation $(x - 2)(x + 1) = 0$, or $x^2 - x - 2 = 0$.

Any integer multiple of this entire equation will also yield the same roots, as the integer will simply cancel out when the equation is factored. For example, $2x^2 - 2x - 4 = 0$ factors as $2(x - 2)(x + 1) = 0$.

Parabolas

A **parabola** is the set of all points in a plane that are equidistant from a fixed line, called the **directrix**, and a fixed point not on the line, called the **focus**. The **axis** is the line perpendicular to the directrix that passes through the focus.

Copyright © Mometrix Media. You have been licensed one copy of this document for personal use only. Any other reproduction or redistribution is strictly prohibited. All rights reserved.
This content is provided for test preparation purposes only and does not imply an endorsement by Mometrix of any particular political, scientific, or religious point of view.

For parabolas that open up or down, the standard equation is $(x - h)^2 = 4c(y - k)$, where h, c, and k are coefficients. If c is positive, the parabola opens up. If c is negative, the parabola opens down. The vertex is the point (h, k). The directrix is the line having the equation $y = -c + k$, and the focus is the point $(h, c + k)$.

For parabolas that open left or right, the standard equation is $(y - k)^2 = 4c(x - h)$, where k, c, and h are coefficients. If c is positive, the parabola opens to the right. If c is negative, the parabola opens to the left. The vertex is the point (h, k). The directrix is the line having the equation $x = -c + h$, and the focus is the point $(c + h, k)$.

Review Video: Parabolas
Visit mometrix.com/academy and enter code: 129187

Review Video: Vertex of a Parabola
Visit mometrix.com/academy and enter code: 272300

Basic Functions

Function and Relation

When expressing functional relationships, the **variables** x and y are typically used. These values are often written as the **coordinates** (x, y). The x-value is the independent variable and the y-value is the dependent variable. A **relation** is a set of data in which there is not a unique y-value for each x-value in the dataset. This means that there can be two of the same x-values assigned to different y-values. A relation is simply a relationship between the x- and y-values in each coordinate but does not apply to the relationship between the values of x and y in the data set. A **function** is a relation where one quantity depends on the other. For example, the amount of money that you make depends on the number of hours that you work. In a function, each x-value in the data set has one unique y-value because the y-value depends on the x-value.

Functions

A function has exactly one value of **output variable** (dependent variable) for each value of the **input variable** (independent variable). The set of all values for the input variable (here assumed to be x) is the domain of the function, and the set of all corresponding values of the output variable (here assumed to be y) is the range of the function. When looking at a graph of an equation, the easiest way to determine if the equation is a function or not is to conduct the vertical line test. If a vertical line drawn through any value of x crosses the graph in more than one place, the equation is not a function.

Determining a Function

You can determine whether an equation is a **function** by substituting different values into the equation for x. You can display and organize these numbers in a data table. A **data table** contains the values for x and y, which you can also list as coordinates. In order for a function to exist, the table cannot contain any repeating x-values that correspond with different y-values. If each x-coordinate has a unique y-coordinate, the table contains a function. However, there can be repeating y-values that correspond with different x-values. An example of this is when the function contains an exponent. Example: if $x^2 = y$, $2^2 = 4$, and $(-2)^2 = 4$.

Review Video: Definition of a Function
Visit mometrix.com/academy and enter code: 784611

Copyright © Mometrix Media. You have been licensed one copy of this document for personal use only. Any other reproduction or redistribution is strictly prohibited. All rights reserved.
This content is provided for test preparation purposes only and does not imply an endorsement by Mometrix of any particular political, scientific, or religious point of view.

Finding the Domain and Range of a Function

The **domain** of a function $f(x)$ is the set of all input values for which the function is defined. The **range** of a function $f(x)$ is the set of all possible output values of the function—that is, of every possible value of $f(x)$, for any value of x in the function's domain. For a function expressed in a table, every input-output pair is given explicitly. To find the domain, we just list all the x-values and to find the range, we just list all the values of $f(x)$. Consider the following example:

x	-1	4	2	1	0	3	8	6
$f(x)$	3	0	3	-1	-1	2	4	6

In this case, the domain would be $\{-1, 4, 2, 1, 0, 3, 8, 6\}$ or, putting them in ascending order, $\{-1, 0, 1, 2, 3, 4, 6, 8\}$. (Putting the values in ascending order isn't strictly necessary, but generally makes the set easier to read.) The range would be $\{3, 0, 3, -1, -1, 2, 4, 6\}$. Note that some of these values appear more than once. This is entirely permissible for a function; while each value of x must be matched to a unique value of $f(x)$, the converse is not true. We don't need to list each value more than once, so eliminating duplicates, the range is $\{3, 0, -1, 2, 4, 6\}$, or, putting them in ascending order, $\{-1, 0, 2, 3, 4, 6\}$.

Note that by definition of a function, no input value can be matched to more than one output value. It is good to double-check to make sure that the data given follows this and is therefore actually a function.

Review Video: Domain and Range
Visit mometrix.com/academy and enter code: 778133

Review Video: Domain and Range of Quadratic Functions
Visit mometrix.com/academy and enter code: 331768

Writing a Function Rule Using a Table

If given a set of data, place the corresponding x- and y-values into a table and analyze the relationship between them. Consider what you can do to each x-value to obtain the corresponding y-value. Try adding or subtracting different numbers to and from x and then try multiplying or dividing different numbers to and from x. If none of these **operations** give you the y-value, try combining the operations. Once you find a rule that works for one pair, make sure to try it with each additional set of ordered pairs in the table. If the same operation or combination of operations satisfies each set of coordinates, then the table contains a function. The rule is then used to write the equation of the function in "$y = f(x)$" form.

Direct and Inverse Variations of Variables

Variables that vary directly are those that either both increase at the same rate or both decrease at the same rate. For example, in the functions $y = kx$ or $y = kx^n$, where k and n are positive, the value of y increases as the value of x increases and decreases as the value of x decreases.

Variables that vary inversely are those where one increases while the other decreases. For example, in the functions $y = \frac{k}{x}$ or $y = \frac{k}{x^n}$ where k and n are positive, the value of y increases as the value of x decreases and decreases as the value of x increases.

In both cases, k is the constant of variation.

Copyright © Mometrix Media. You have been licensed one copy of this document for personal use only. Any other reproduction or redistribution is strictly prohibited. All rights reserved.
This content is provided for test preparation purposes only and does not imply an endorsement by Mometrix of any particular political, scientific, or religious point of view.

PROPERTIES OF FUNCTIONS

There are many different ways to classify functions based on their structure or behavior. Important features of functions include:

- **End behavior**: the behavior of the function at extreme values ($f(x)$ as $x \to \pm\infty$)
- **y-intercept**: the value of the function at $f(0)$
- **Roots**: the values of x where the function equals zero ($f(x) = 0$)
- **Extrema**: minimum or maximum values of the function or where the function changes direction ($f(x) \geq k$ or $f(x) \leq k$)

CLASSIFICATION OF FUNCTIONS

An **invertible function** is defined as a function, $f(x)$, for which there is another function, $f^{-1}(x)$, such that $f^{-1}(f(x)) = x$. For example, if $f(x) = 3x - 2$ the inverse function, $f^{-1}(x)$, can be found:

$$x = 3(f^{-1}(x)) - 2$$
$$\frac{x+2}{3} = f^{-1}(x)$$

$$f^{-1}(f(x)) = \frac{3x - 2 + 2}{3}$$
$$= \frac{3x}{3}$$
$$= x$$

Note that $f^{-1}(x)$ is a valid function over all values of x.

In a **one-to-one function**, each value of x has exactly one value for y on the coordinate plane (this is the definition of a function) and each value of y has exactly one value for x. While the vertical line test will determine if a graph is that of a function, the horizontal line test will determine if a function is a one-to-one function. If a horizontal line drawn at any value of y intersects the graph in more than one place, the graph is not that of a one-to-one function. Do not make the mistake of using the horizontal line test exclusively in determining if a graph is that of a one-to-one function. A one-to-one function must pass both the vertical line test and the horizontal line test. As such, one-to-one functions are invertible functions.

A **many-to-one function** is a function whereby the relation is a function, but the inverse of the function is not a function. In other words, each element in the domain is mapped to one and only one element in the range. However, one or more elements in the range may be mapped to the same element in the domain. A graph of a many-to-one function would pass the vertical line test, but not the horizontal line test. This is why many-to-one functions are not invertible.

A **monotone function** is a function whose graph either constantly increases or constantly decreases. Examples include the functions $f(x) = x$, $f(x) = -x$, or $f(x) = x^3$.

An **even function** has a graph that is symmetric with respect to the y-axis and satisfies the equation $f(x) = f(-x)$. Examples include the functions $f(x) = x^2$ and $f(x) = ax^n$, where a is any real number and n is a positive even integer.

An **odd function** has a graph that is symmetric with respect to the origin and satisfies the equation $f(x) = -f(-x)$. Examples include the functions $f(x) = x^3$ and $f(x) = ax^n$, where a is any real number and n is a positive odd integer.

Review Video: Even and Odd Functions
Visit mometrix.com/academy and enter code: 278985

Copyright © Mometrix Media. You have been licensed one copy of this document for personal use only. Any other reproduction or redistribution is strictly prohibited. All rights reserved.
This content is provided for test preparation purposes only and does not imply an endorsement by Mometrix of any particular political, scientific, or religious point of view.

Constant functions are given by the equation $f(x) = b$, where b is a real number. There is no independent variable present in the equation, so the function has a constant value for all x. The graph of a constant function is a horizontal line of slope 0 that is positioned b units from the x-axis. If b is positive, the line is above the x-axis; if b is negative, the line is below the x-axis.

Identity functions are identified by the equation $f(x) = x$, where every value of the function is equal to its corresponding value of x. The only zero is the point (0,0). The graph is a line with a slope of 1.

In **linear functions**, the value of the function changes in direct proportion to x. The rate of change, represented by the slope on its graph, is constant throughout. The standard form of a linear equation is $ax + cy = d$, where a, c, and d are real numbers. As a function, this equation is commonly in the form $y = mx + b$ or $f(x) = mx + b$ where $m = -\frac{a}{c}$ and $b = \frac{d}{c}$. This is known as the slope-intercept form, because the coefficients give the slope of the graphed function (m) and its y-intercept (b). Solve the equation $mx + b = 0$ for x to get $x = -\frac{b}{m}$, which is the only zero of the function. The domain and range are both the set of all real numbers.

Review Video: Graphing Linear Functions
Visit mometrix.com/academy and enter code: 699478

Algebraic functions are those that exclusively use polynomials and roots. These would include polynomial functions, rational functions, square root functions, and all combinations of these functions, such as polynomials as the radicand. These combinations may be joined by addition, subtraction, multiplication, or division, but may not include variables as exponents.

Review Video: Common Functions
Visit mometrix.com/academy and enter code: 629798

Absolute Value Functions

An **absolute value function** is in the format $f(x) = |ax + b|$. Like other functions, the domain is the set of all real numbers. However, because absolute value indicates positive numbers, the range is limited to positive real numbers. To find the zero of an absolute value function, set the portion inside the absolute value sign equal to zero and solve for x. An absolute value function is also known as a piecewise function because it must be solved in pieces—one for if the value inside the absolute value sign is positive, and one for if the value is negative. The function can be expressed as:

$$f(x) = \begin{cases} ax + b \text{ if } ax + b \geq 0 \\ -(ax + b) \text{ if } ax + b < 0 \end{cases}$$

Copyright © Mometrix Media. You have been licensed one copy of this document for personal use only. Any other reproduction or redistribution is strictly prohibited. All rights reserved.
This content is provided for test preparation purposes only and does not imply an endorsement by Mometrix of any particular political, scientific, or religious point of view.

This will allow for an accurate statement of the range. The graph of an example absolute value function, $f(x) = |2x - 1|$, is below:

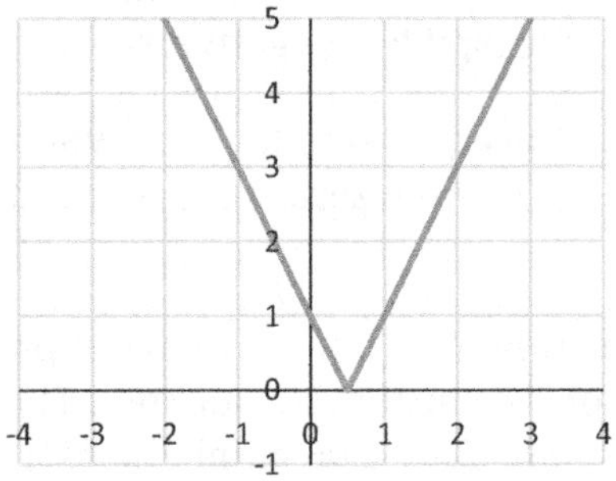

Piecewise Functions

A **piecewise function** is a function that has different definitions on two or more different intervals. The following, for instance, is one example of a piecewise-defined function:

$$f(x) = \begin{cases} x^2, & x < 0 \\ x, & 0 \le x \le 2 \\ (x-2)^2, & x > 2 \end{cases}$$

To graph this function, you would simply graph each part separately in the appropriate domain. The final graph would look like this:

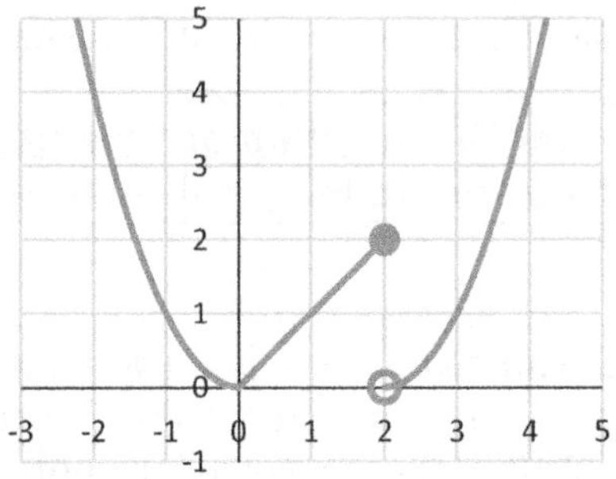

Note the filled and hollow dots at the discontinuity at $x = 2$. This is important to show which side of the graph that point corresponds to. Because $f(x) = x$ on the closed interval $0 \le x \le 2$, $f(2) = 2$. The point $(2, 2)$ is therefore marked with a filled circle, and the point (2,0), which is the endpoint of the rightmost $(x - 2)^2$ part of the graph but *not actually part of the function*, is marked with a hollow dot to indicate this.

Review Video: Piecewise Functions
Visit mometrix.com/academy and enter code: 707921

Quadratic Functions

A **quadratic function** is a function in the form $y = ax^2 + bx + c$, where a does not equal 0. While a linear function forms a line, a quadratic function forms a **parabola**, which is a u-shaped figure that either opens upward or downward. A parabola that opens upward is said to be a **positive quadratic function,** and a parabola that opens downward is said to be a **negative quadratic function**. The shape of a parabola can differ, depending on the values of a, b, and c. All parabolas contain a **vertex**, which is the highest possible point, the **maximum**, or the lowest possible point, the **minimum**. This is the point where the graph begins moving in the opposite direction. A

Copyright © Mometrix Media. You have been licensed one copy of this document for personal use only. Any other reproduction or redistribution is strictly prohibited. All rights reserved.
This content is provided for test preparation purposes only and does not imply an endorsement by Mometrix of any particular political, scientific, or religious point of view.

quadratic function can have zero, one, or two solutions, and therefore zero, one, or two x-intercepts. Recall that the x-intercepts are referred to as the zeros, or roots, of a function. A quadratic function will have only one y-intercept. Understanding the basic components of a quadratic function can give you an idea of the shape of its graph.

Example graph of a positive quadratic function, $x^2 + 2x - 3$:

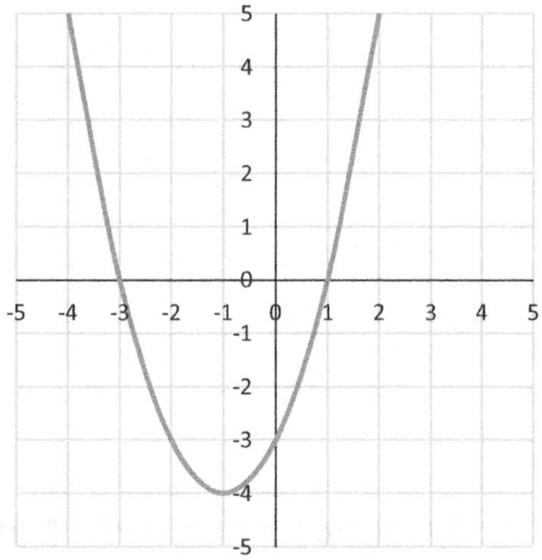

POLYNOMIAL FUNCTIONS

A **polynomial function** is a function with multiple terms and multiple powers of x, such as:

$$f(x) = a_n x^n + a_{n-1}x^{n-1} + a_{n-2}x^{n-2} + \cdots + a_1 x + a_0$$

where n is a non-negative integer that is the highest exponent in the polynomial and $a_n \neq 0$. The domain of a polynomial function is the set of all real numbers. If the greatest exponent in the polynomial is even, the polynomial is said to be of even degree and the range is the set of real numbers that satisfy the function. If the greatest exponent in the polynomial is odd, the polynomial is said to be odd and the range, like the domain, is the set of all real numbers.

RATIONAL FUNCTIONS

A **rational function** is a function that can be constructed as a ratio of two polynomial expressions: $f(x) = \frac{p(x)}{q(x)}$, where $p(x)$ and $q(x)$ are both polynomial expressions and $q(x) \neq 0$. The domain is the set of all real numbers, except any values for which $q(x) = 0$. The range is the set of real numbers that satisfies the function when the domain is applied. When you graph a rational function, you will have vertical asymptotes wherever $q(x) = 0$. If the polynomial in the numerator is of lesser degree than the polynomial in the denominator, the x-axis will also be a horizontal asymptote. If the numerator and denominator have equal degrees, there will be a horizontal asymptote not on the x-axis. If the degree of the numerator is exactly one greater than the degree of the denominator, the graph will have an oblique, or diagonal, asymptote. The asymptote will be along the line $y = \frac{p_n}{q_{n-1}}x + \frac{p_{n-1}}{q_{n-1}}$, where p_n and q_{n-1} are the coefficients of the highest degree terms in their respective polynomials.

SQUARE ROOT FUNCTIONS

A **square root function** is a function that contains a radical and is in the format $f(x) = \sqrt{ax + b}$. The domain is the set of all real numbers that yields a positive radicand or a radicand equal to zero.

Copyright © Mometrix Media. You have been licensed one copy of this document for personal use only. Any other reproduction or redistribution is strictly prohibited. All rights reserved.
This content is provided for test preparation purposes only and does not imply an endorsement by Mometrix of any particular political, scientific, or religious point of view.

Because square root values are assumed to be positive unless otherwise identified, the range is all real numbers from zero to infinity. To find the zero of a square root function, set the radicand equal to zero and solve for x. The graph of a square root function is always to the right of the zero and always above the x-axis.

Example graph of a square root function, $f(x) = \sqrt{2x + 1}$:

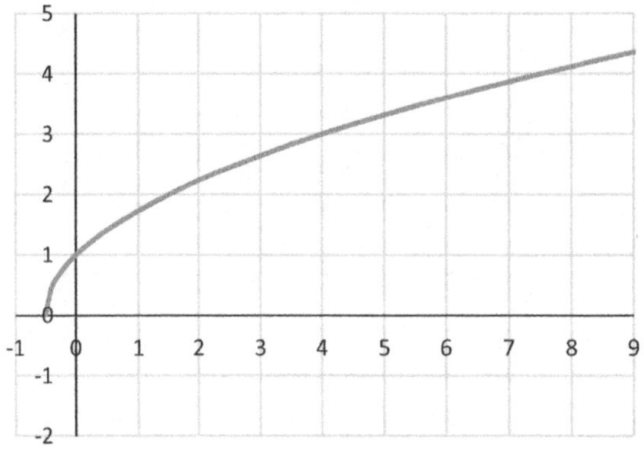

Basic Trigonometric Functions

SINE

The **sine** (sin) function has a period of 360° or 2π radians. This means that its graph makes one complete cycle every 360° or 2π. Because $\sin 0 = 0$, the graph of $y = \sin x$ begins at the origin, with the x-axis representing the angle measure, and the y-axis representing the sine of the angle. The graph of the sine function is a smooth curve that begins at the origin, peaks at the point $\left(\frac{\pi}{2}, 1\right)$, crosses the x-axis at $(\pi, 0)$, has its lowest point at $\left(\frac{3\pi}{2}, -1\right)$, and returns to the x-axis to complete one cycle at $(2\pi, 0)$.

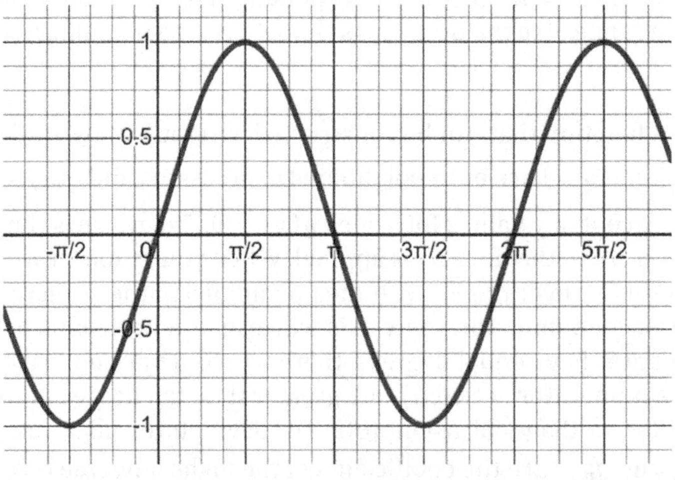

Review Video: Sine
Visit mometrix.com/academy and enter code: 193339

COSINE

The **cosine** (cos) function also has a period of 360° or 2π radians, which means that its graph also makes one complete cycle every 360° or 2π. Because $\cos 0° = 1$, the graph of $y = \cos x$ begins at the

Copyright © Mometrix Media. You have been licensed one copy of this document for personal use only. Any other reproduction or redistribution is strictly prohibited. All rights reserved.
This content is provided for test preparation purposes only and does not imply an endorsement by Mometrix of any particular political, scientific, or religious point of view.

point $(0, 1)$, with the x-axis representing the angle measure, and the y-axis representing the cosine of the angle. The graph of the cosine function is a smooth curve that begins at the point $(0,1)$, crosses the x-axis at the point $\left(\frac{\pi}{2}, 0\right)$, has its lowest point at $(\pi, -1)$, crosses the x-axis again at the point $\left(\frac{3\pi}{2}, 0\right)$, and returns to a peak at the point $(2\pi, 1)$ to complete one cycle.

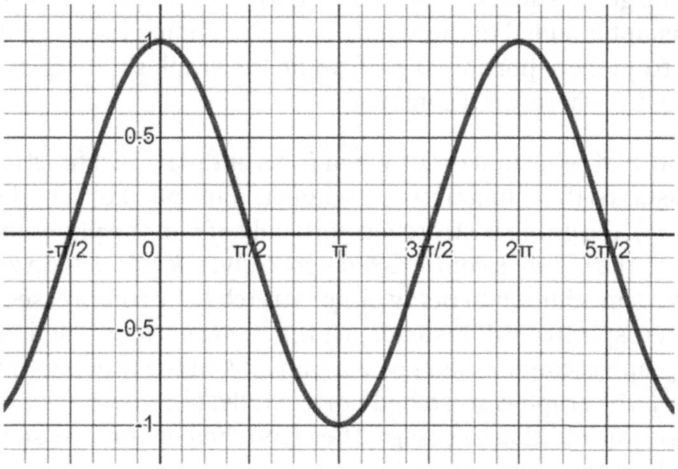

Review Video: Cosine
Visit mometrix.com/academy and enter code: 361120

TANGENT

The **tangent** (tan) function has a period of 180° or π radians, which means that its graph makes one complete cycle every 180° or π radians. The x-axis represents the angle measure, and the y-axis represents the tangent of the angle. The graph of the tangent function is a series of smooth curves that cross the x-axis at every 180° or π radians and have an asymptote every $k \times 90°$ or $\frac{k\pi}{2}$ radians, where k is an odd integer. This can be explained by the fact that the tangent is calculated by dividing the sine by the cosine, since the cosine equals zero at those asymptote points.

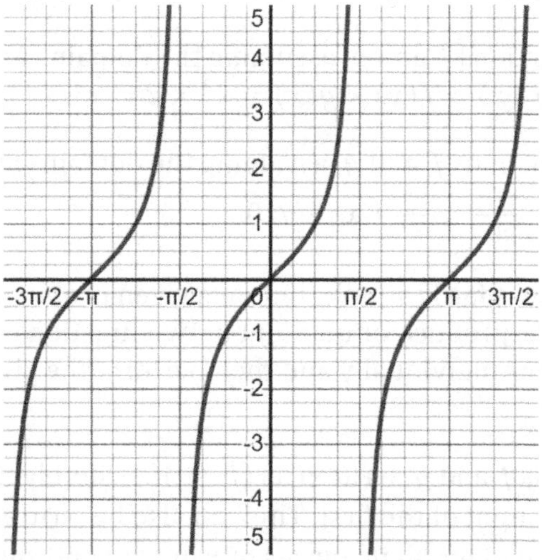

Review Video: Tangent
Visit mometrix.com/academy and enter code: 947639

Copyright © Mometrix Media. You have been licensed one copy of this document for personal use only. Any other reproduction or redistribution is strictly prohibited. All rights reserved.
This content is provided for test preparation purposes only and does not imply an endorsement by Mometrix of any particular political, scientific, or religious point of view.

Working with Functions

Manipulation of Functions

Translation occurs when values are added to or subtracted from the x- or y-values. If a constant is added to the y-portion of each point, the graph shifts up. If a constant is subtracted from the y-portion of each point, the graph shifts down. This is represented by the expression $f(x) \pm k$, where k is a constant. If a constant is added to the x-portion of each point, the graph shifts left. If a constant is subtracted from the x-portion of each point, the graph shifts right. This is represented by the expression $f(x \pm k)$, where k is a constant.

Stretching, compression, and reflection occur when different parts of a function are multiplied by different groups of constants. If the function as a whole is multiplied by a real number constant greater than 1, $(k \times f(x))$, the graph is stretched vertically. If k in the previous equation is greater than zero but less than 1, the graph is compressed vertically. If k is less than zero, the graph is reflected about the x-axis, in addition to being either stretched or compressed vertically if k is less than or greater than –1, respectively. If instead, just the x-term is multiplied by a constant greater than 1 $(f(k \times x))$, the graph is compressed horizontally. If k in the previous equation is greater than zero but less than 1, the graph is stretched horizontally. If k is less than zero, the graph is reflected about the y-axis, in addition to being either stretched or compressed horizontally if k is greater than or less than –1, respectively.

Review Video: Manipulation of Functions
Visit mometrix.com/academy and enter code: 669117

Applying the Basic Operations to Functions

For each of the basic operations, we will use these functions as examples: $f(x) = x^2$ and $g(x) = x$.

To find the sum of two functions f and g, assuming the domains are compatible, simply add the two functions together: $(f + g)(x) = f(x) + g(x) = x^2 + x$.

To find the difference of two functions f and g, assuming the domains are compatible, simply subtract the second function from the first: $(f - g)(x) = f(x) - g(x) = x^2 - x$.

To find the product of two functions f and g, assuming the domains are compatible, multiply the two functions together: $(f \times g)(x) = f(x) \times g(x) = x^2 \times x = x^3$.

To find the quotient of two functions f and g, assuming the domains are compatible, divide the first function by the second: $\frac{f}{g}(x) = \frac{f(x)}{g(x)} = \frac{x^2}{x} = x\,; x \neq 0$.

The example given in each case is fairly simple, but on a given problem, if you are looking only for the value of the sum, difference, product, or quotient of two functions at a particular x-value, it may be simpler to solve the functions individually and then perform the given operation using those values.

The composite of two functions f and g, written as $(f \circ g)(x)$ simply means that the output of the second function is used as the input of the first. This can also be written as $f(g(x))$. In general, this can be solved by substituting $g(x)$ for all instances of x in $f(x)$ and simplifying. Using the example functions $f(x) = x^2 - x + 2$ and $g(x) = x + 1$, we can find that $(f \circ g)(x)$ or $f(g(x))$ is equal to $f(x + 1) = (x + 1)^2 - (x + 1) + 2$, which simplifies to $x^2 + x + 2$.

Copyright © Mometrix Media. You have been licensed one copy of this document for personal use only. Any other reproduction or redistribution is strictly prohibited. All rights reserved.
This content is provided for test preparation purposes only and does not imply an endorsement by Mometrix of any particular political, scientific, or religious point of view.

It is important to note that $(f \circ g)(x)$ is not necessarily the same as $(g \circ f)(x)$. The process is not always commutative like addition or multiplication expressions. It *can* be commutative, but most often this is not the case.

EVALUATING LINEAR FUNCTIONS

A **function** can be expressed as an equation that relates an input to an output where each input corresponds to exactly one output. The input of a function is defined by the x-variable, and the output is defined by the y-variable. For example, consider the function $y = 2x + 6$. The value of y, the output, is determined by the value of the x, the input. If the value of x is 3, the value of y is $y = 2(3) + 6 = 6 + 6 = 12$. This means that when $x = 3$, $y = 12$. This can be expressed as the ordered pair $(3,12)$.

It is common for function equations to use the form $f(x) =$ instead of $y =$. However, $f(x)$ and y represent the same thing. We read $f(x)$ as "f of x." "f of x" implies that the value of f depends on the value of x. The function used in the example above could be expressed as $y = 2x + 6$ or $f(x) = 2x + 6$. Both functions represent the same line when graphed.

Functions that are expressed in the form $f(x) =$ are evaluated in the same way the equations are evaluated in the form $y =$. For example, when evaluating the function $f(x) = 3x - 2$ for $f(6)$, substitute 6 in for x, and simplify. $f(x) = 3x - 2$ becomes $f(6) = 3(6) - 2 = 18 - 2 = 16$. When x is 6, $f(x)$ is 16.

Find the value of $f(8)$.

$$f(x) = 3x - 2$$
$$f(8) = 3(8) - 2$$
$$f(8) = 22$$

Advanced Functions

STEP FUNCTIONS

The double brackets indicate a step function. For a step function, the value inside the double brackets is rounded down to the nearest integer. The graph of the function $f_0(x) = [\![x]\!]$ appears on the left graph. In comparison $f(x) = 2\left[\!\!\left[\frac{1}{3}(x-1)\right]\!\!\right]$ is on the right graph. The coefficient of 2 shows that it's stretched vertically by a factor of 2 (so there's a vertical distance of 2 units between successive "steps"). The coefficient of $\frac{1}{3}$ in front of the x shows that it's stretched horizontally by a

Copyright © Mometrix Media. You have been licensed one copy of this document for personal use only. Any other reproduction or redistribution is strictly prohibited. All rights reserved.
This content is provided for test preparation purposes only and does not imply an endorsement by Mometrix of any particular political, scientific, or religious point of view.

factor of 3 (so each "step" is three units long), and the $x - 1$ shows that it's displaced one unit to the right.

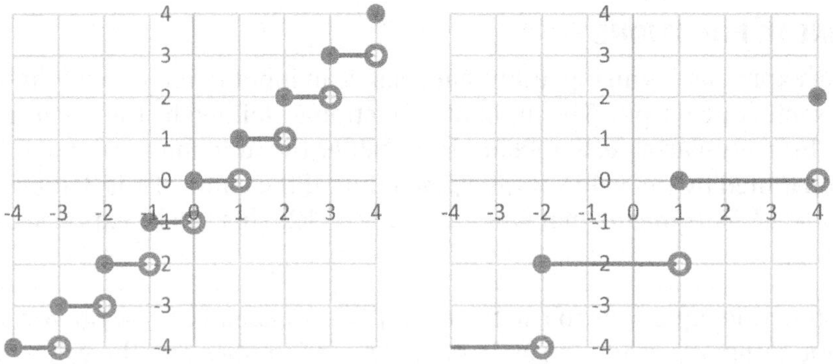

TRANSCENDENTAL FUNCTIONS

Transcendental functions are all functions that are non-algebraic. Any function that includes logarithms, trigonometric functions, variables as exponents, or any combination that includes any of these is not algebraic in nature, even if the function includes polynomials or roots.

EXPONENTIAL FUNCTIONS

Exponential functions are equations that have the format $y = b^x$, where base $b > 0$ and $b \neq 1$. The exponential function can also be written $f(x) = b^x$. Recall the properties of exponents, like the product of terms with the same base is equal to the base raised to the sum of the exponents ($a^x \times a^y = a^{x+y}$) and a term with an exponent that is raised to an exponent is equal to the base of the original term raised to the product of the exponents: ($(a^x)^y = a^{xy}$). The graph of an example exponential function, $f(x) = 2^x$, is below:

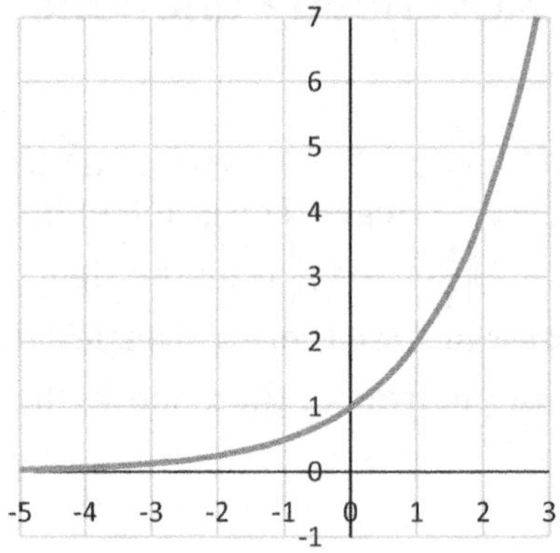

Note in the graph that the y-value approaches zero to the left and infinity to the right. One of the key features of an exponential function is that there will be one end that goes off to infinity and another that asymptotically approaches a lower bound. Common forms of exponential functions include:

Geometric sequences: $a_n = a_1 \times r^{n-1}$, where a_n is the value of the nth term, a_1 is the initial value, r is the common ratio, and n is the number of terms. Note that $a_1 \times r^{1-1} = a_1 \times r^0 = a_1 \times 1 = a_1$.

Copyright © Mometrix Media. You have been licensed one copy of this document for personal use only. Any other reproduction or redistribution is strictly prohibited. All rights reserved.
This content is provided for test preparation purposes only and does not imply an endorsement by Mometrix of any particular political, scientific, or religious point of view.

Population growth: $f(t) = ae^{rt}$, where $f(t)$ is the population at time $t \geq 0$, a is the initial population, e is the mathematical constant known as Euler's number, and r is the growth rate.

Review Video: Population Growth
Visit mometrix.com/academy and enter code: 109278

Compound interest: $f(t) = P\left(1 + \frac{r}{n}\right)^{nt}$, where $f(t)$ is the account value at a certain number of time periods $t \geq 0$, P is the initial principal balance, r is the interest rate, and n is the number of times the interest is applied per time period.

Review Video: Interest Functions
Visit mometrix.com/academy and enter code: 559176

General exponential growth or decay: $f(t) = a(1 + r)^t$, where $f(t)$ is the future count, a is the current or initial count, r is the growth or decay rate, and t is the time.

For example, suppose the initial population of a town was 1,200 people. The annual population growth is 5%. The current population is 2,400. To find out how much time has passed since the town was founded, we can use the following function:

$$2{,}400 = 1{,}200e^{0.05t}.$$

The general form for population growth may be represented as $f(t) = ae^{rt}$, where $f(t)$ represents the current population, a represents the initial population, r represents the growth rate, and t represents the time. Thus, substituting the initial population, current population, and rate into this form gives the equation above.

The number of years that have passed were found by first dividing both sides of the equation by 1,200. Doing so gives $2 = e^{0.05t}$. Taking the natural logarithm of both sides gives $\ln(2) = \ln(e^{0.05t})$. Applying the power property of logarithms, the equation may be rewritten as $\ln(2) = 0.05t \times \ln(e)$, which simplifies as $\ln(2) = 0.05t$. Dividing both sides of this equation by 0.05 gives $t \approx 13.86$. Thus, approximately 13.86 years passed.

Logarithmic Functions

Logarithmic functions are equations that have the format $y = \log_b x$ or $f(x) = \log_b x$. The base b may be any number except one; however, the most common bases for logarithms are base 10 and base e. The log base e is the natural logarithm, or ln, expressed by the function $f(x) = \ln x$.

Any logarithm that does not have an assigned value of b is assumed to be base 10: $\log x = \log_{10} x$. Exponential functions and logarithmic functions are related in that one is the inverse of the other. If $f(x) = b^x$, then $f^{-1}(x) = \log_b x$. This can perhaps be expressed more clearly by the two equations: $y = b^x$ and $x = \log_b y$.

Copyright © Mometrix Media. You have been licensed one copy of this document for personal use only. Any other reproduction or redistribution is strictly prohibited. All rights reserved.
This content is provided for test preparation purposes only and does not imply an endorsement by Mometrix of any particular political, scientific, or religious point of view.

The following properties apply to logarithmic expressions:

Property	Description
$\log_b 1 = 0$	The log of 1 is equal to 0 for any base
$\log_b b = 1$	The log of the base is equal to 1
$\log_b b^p = p$	The log of the base raised to a power is equal to that power
$\log_b MN = \log_b M + \log_b N$	The log of a product is the sum of the log of each factor
$\log_b \frac{M}{N} = \log_b M - \log_b N$	The log of a quotient is equal to the log of the dividend minus the log of the divisor
$\log_b M^p = p \log_b M$	The log of a value raised to a power is equal to the power times the log of the value

The graph of an example logarithmic function, $f(x) = \log_2(x + 2)$, is below:

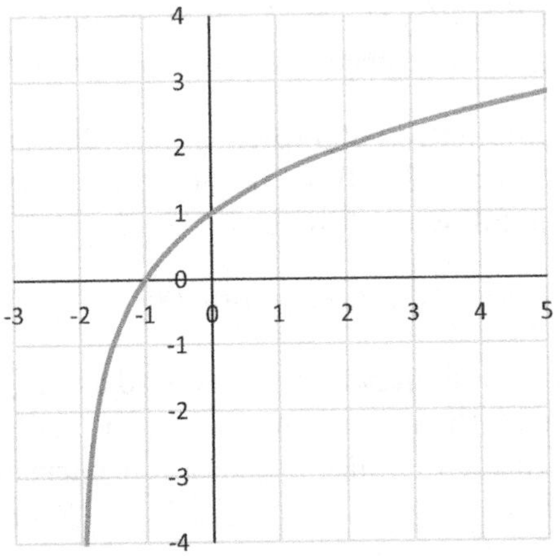

Review Video: Logarithmic Function
Visit mometrix.com/academy and enter code: 658985

Trigonometric Functions

Trigonometric functions are periodic, meaning that they repeat the same form over and over. The basic trigonometric functions are sine (abbreviated 'sin'), cosine (abbreviated 'cos'), and tangent (abbreviated 'tan'). The simplest way to think of them is as describing the ratio of the side lengths of a right triangle in relation to the angles of the triangle.

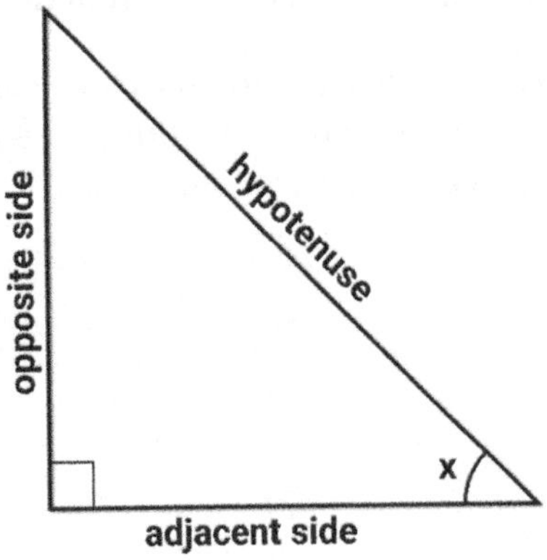

$$\sin x = \frac{\text{opposite side}}{\text{hypotenuse}}$$
$$\cos x = \frac{\text{adjacent side}}{\text{hypotenuse}}$$
$$\tan x = \frac{\text{opposite side}}{\text{adjacent side}}$$

Copyright © Mometrix Media. You have been licensed one copy of this document for personal use only. Any other reproduction or redistribution is strictly prohibited. All rights reserved.
This content is provided for test preparation purposes only and does not imply an endorsement by Mometrix of any particular political, scientific, or religious point of view.

Using sine as an example, trigonometric functions take the form $f(x) = A\sin(Bx + C) + D$, where the **amplitude** is simply equal to A. The **period** is the distance between successive peaks or troughs, essentially the length of the repeated pattern. In this form, the period is equal to $\frac{2\pi}{B}$. As for C, this is the **phase shift** or the horizontal shift of the function. The last term, D, is the vertical shift and determines the **midline** as $y = D$.

For instance, consider the function $f(x) = 2 + \frac{3}{2}\sin\left(\pi x + \frac{\pi}{2}\right)$. Here, $A = \frac{3}{2}$, $B = \pi$, $C = \frac{\pi}{2}$, and $D = 2$, so the midline is at $y = 2$, the amplitude is $\frac{3}{2}$, and the period is $\frac{2\pi}{\pi} = 2$. To graph this function, we center the sine wave on the midline and extend it to a height above and below the midline equal to the amplitude—so this graph would have a minimum value of $2 - \frac{3}{2} = \frac{1}{2}$ and a maximum of $2 + \frac{3}{2} = \frac{7}{2}$. So, the function would be graphed as follows:

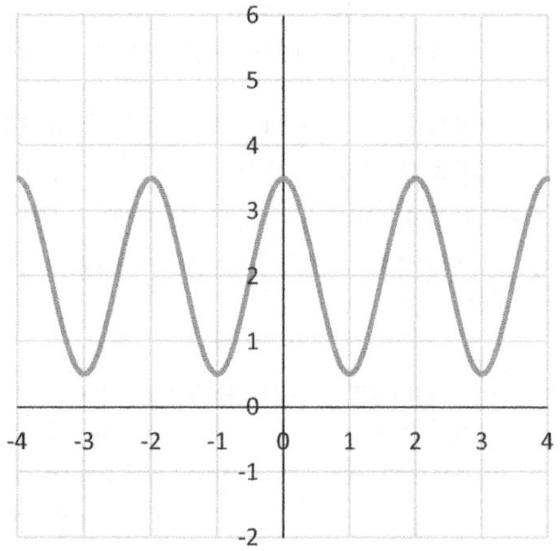

Complex Numbers

Complex numbers consist of a real component and an imaginary component. Complex numbers are expressed in the form $a + bi$ with real component a and imaginary component bi. The imaginary unit i is equal to $\sqrt{-1}$. That means $i^2 = -1$. The imaginary unit provides a way to find the square root of a negative number. For example, $\sqrt{-25}$ is $5i$. You should expect questions asking you to add, subtract, multiply, divide, and simplify complex numbers. You may see a question that says, "Add $3 + 2i$ and $5 - 7i$" or "Subtract $4 + i\sqrt{5}$ from $2 + i\sqrt{5}$." Or you may see a question that says, "Multiply $6 + 2i$ by $8 - 4i$" or "Divide $1 - 3i$ by $9 - 7i$."

Operations on Complex Numbers

Operations with complex numbers resemble operations with variables in algebra. When adding or subtracting complex numbers, you can only combine like terms—real terms with real terms and imaginary terms with imaginary terms. For example, if you are asked to simplify the expression $-2 + 4i - (-3 + 7i) - 5i$, you should first remove the parentheses to yield $-2 + 4i + 3 - 7i - 5i$. Combining like terms yields $1 - 8i$. One interesting aspect of imaginary numbers is that if i has an exponent greater than 1, it can be simplified. Example: $i^2 = -1$, $i^3 = -i$, and $i^4 = 1$. When multiplying complex numbers, remember to simplify each i with an exponent greater than 1. For example, you might see a question that says, "Simplify $(2 - i)(3 + 2i)$." You need to distribute and multiply to get $6 + 4i - 3i - 2i^2$. This is further simplified to $6 + i - 2(-1)$, or $8 + i$.

Copyright © Mometrix Media. You have been licensed one copy of this document for personal use only. Any other reproduction or redistribution is strictly prohibited. All rights reserved.
This content is provided for test preparation purposes only and does not imply an endorsement by Mometrix of any particular political, scientific, or religious point of view.

SIMPLIFYING EXPRESSIONS WITH COMPLEX DENOMINATORS

If an expression contains an i in the denominator, it must be simplified. Remember, roots cannot be left in the denominator of a fraction. Since i is equivalent to $\sqrt{-1}$, i cannot be left in the denominator of a fraction. You must rationalize the denominator of a fraction that contains a complex denominator by multiplying the numerator and denominator by the conjugate of the denominator. The conjugate of the complex number $a + bi$ is $a - bi$. You can simplify $\frac{2}{5i}$ by simply multiplying $\frac{2}{5i} \times \frac{i}{i}$, which yields $-\frac{2}{5}i$. And you can simplify $\frac{5+3i}{2-4i}$ by multiplying $\frac{5+3i}{2-4i} \times \frac{2+4i}{2+4i}$. This yields $\frac{10+20i+6i-12}{4-8i+8i+16}$ which simplifies to $\frac{-2+26i}{20}$ or $\frac{-1+13i}{10}$, which can also be written as $-\frac{1}{10} + \frac{13}{10}i$.

Sequences

A **sequence** is an ordered set of numbers that continues in a defined pattern. The function that defines a sequence has a domain composed of the set of positive integers. Each member of the sequence is an element, or individual term. Each element is identified by the notation a_n, where a is the term of the sequence, and n is the integer identifying which term in the sequence a is.

There are two different ways to represent a sequence that contains the element a_n. The first is the simple notation $\{a_n\}$. The second is the expanded notation of a sequence: $a_1, a_2, a_3, \dots a_n, \dots$. Notice that the expanded form does not end with the nth term. There is no indication that the nth term is the last term in the sequence, only that the nth term is an element of the sequence.

ARITHMETIC SEQUENCES

An **arithmetic sequence**, or arithmetic progression, is a special kind of sequence in which a specific quantity, called the common difference, is added to each term to make the next term. The common difference may be positive or negative. The general form of an arithmetic sequence containing n terms is $a_1, a_1 + d, a_1 + 2d, \dots, a_1 + (n-1)d$, where d is the common difference. The general formula for any term of an arithmetic sequence is $a_n = a_1 + (n-1)d$, where a_n is the term you are looking for and d is the common difference. To find the sum of the first n terms of an arithmetic sequence, use the formula $s_n = \frac{n}{2}(a_1 + a_n)$.

Review Video: Arithmetic Sequence
Visit mometrix.com/academy and enter code: 676885

MONOTONIC SEQUENCES

A **monotonic sequence** is a sequence that is either nonincreasing or nondecreasing. A **nonincreasing** sequence is one whose terms either get progressively smaller in value or remain the same. Such a sequence is always bounded above, that is, all elements of the sequence must be less than some real number. A **nondecreasing** sequence is one whose terms either get progressively larger in value or remain the same. Such a sequence is always bounded below, that is, all elements of the sequence must be greater than some real number.

RECURSIVE SEQUENCES

When one element of a sequence is defined in terms of a previous element or elements of the sequence, the sequence is a **recursive sequence**. For example, given the recursive definition $a_1 = 1; a_2 = 1; a_n = a_{n-1} + a_{n-2}$ for all $n > 2$, you get the sequence 1,1,2,3,5,8, ... This is known as the Fibonacci sequence: a continuing sequence of numbers in which each number (after a_2) is the sum of the two previous numbers. The Fibonacci sequence can be defined as starting with either 1,1

Copyright © Mometrix Media. You have been licensed one copy of this document for personal use only. Any other reproduction or redistribution is strictly prohibited. All rights reserved.
This content is provided for test preparation purposes only and does not imply an endorsement by Mometrix of any particular political, scientific, or religious point of view.

or 0,1. Both definitions are considered correct in mathematics. Make sure you know which definition you are working with when dealing with Fibonacci numbers.

Sometimes in a recursive sequence, the terms can be found using a general formula that does not involve the previous terms of the sequence. Such a formula is called a **closed-form** expression for a recursive definition—an alternate formula that will generate the same sequence of numbers. However, not all sequences based on recursive definitions will have a closed-form expression. Some sequences will require the use of the recursive definition.

The Golden Ratio and the Fibonacci Sequence

The golden ratio is approximately 1.6180339887 and is often represented by the Greek letter phi, Φ. The exact value of Φ is $\frac{(1+\sqrt{5})}{2}$ and it is one of the solutions to $x - \frac{1}{x} = 1$. The golden ratio can be found using the Fibonacci sequence, since the ratio of a term to the previous term approaches Φ as the sequence approaches infinity:

n	a_n	a_{n-1}	$\frac{a_n}{a_{n-1}}$
3	2	1	2
4	3	2	1.5
5	5	3	$1.\overline{6}$
6	8	5	1.6
7	13	8	1.625
8	21	13	$1.\overline{615384}$
9	34	21	$1.\overline{619047}$
⋮	⋮	⋮	⋮
20	6,765	4,181	1.618033963 ...

Geometric Sequences

A geometric sequence is a sequence in which each term is multiplied by a constant number (called the common ratio) to get the next term. Essentially, it's the same concept as an arithmetic sequence, but with multiplication instead of addition.

Consider the following example of a geometric sequence: Andy opens a savings account with $10. During each subsequent week, he plans to double the amount from the previous week.

Sequence: $10, 20, 40, 80, 160, \ldots$

Function: $a_n = 10 \times 2^{n-1}$

This is a geometric sequence with a common ratio of 2. All geometric sequences represent exponential functions. The nth term in any geometric sequence is $a_n = a_1 \times r^{n-1}$, where a_n represents the value of the nth term, a_1 is the initial term, r is the common ratio, and n is the number of terms. Thus, substituting the initial value of 10 and common ratio of 2 gives the function $a_n = 10 \times 2^{n-1}$.

Review Video: Geometric Sequences
Visit mometrix.com/academy and enter code: 140779

Copyright © Mometrix Media. You have been licensed one copy of this document for personal use only. Any other reproduction or redistribution is strictly prohibited. All rights reserved.
This content is provided for test preparation purposes only and does not imply an endorsement by Mometrix of any particular political, scientific, or religious point of view.

Advanced Sequences and Series

Limit of a Sequence

Some sequences will have a **limit**—a value the sequence approaches, or sometimes even reaches, but never passes. A sequence with a limit is called a **convergent** sequence because all the values of the sequence seemingly converge at that point. Sequences that do not converge at a particular limit are **divergent** sequences. The easiest way to determine whether a sequence converges or diverges is to find the limit of the sequence. If the limit is a real number, the sequence is convergent. If the limit is infinity, the sequence is divergent.

Remember the following rules for finding limits:

- $\lim_{n\to\infty} k = k$, for all real numbers k
- $\lim_{n\to\infty} \frac{1}{n} = 0$
- $\lim_{n\to\infty} n = \infty$
- $\lim_{n\to\infty} \frac{k}{n^p} = 0$, for all real numbers k and positive rational numbers p
- The limit of the sum of two sequences is equal to the sum of the limits of the two sequences: $\lim_{n\to\infty} (a_n + b_n) = \lim_{n\to\infty} a_n + \lim_{n\to\infty} b_n$
- The limit of the difference between two sequences is equal to the difference between the limits of the two sequences: $\lim_{n\to\infty} (a_n - b_n) = \lim_{n\to\infty} a_n - \lim_{n\to\infty} b_n$
- The limit of the product of two sequences is equal to the product of the limits of the two sequences: $\lim_{n\to\infty} (a_n \times b_n) = \lim_{n\to\infty} a_n \times \lim_{n\to\infty} b_n$
- The limit of the quotient of two sequences is equal to the quotient of the limits of the two sequences, with some exceptions: $\lim_{n\to\infty} \left(\frac{a_n}{b_n}\right) = \frac{\lim_{n\to\infty} a_n}{\lim_{n\to\infty} b_n}$. In the quotient formula, it is important that $b_n \neq 0$ and that $\lim_{n\to\infty} b_n \neq 0$.
- The limit of a sequence multiplied by a scalar is equal to the scalar multiplied by the limit of the sequence: $\lim_{n\to\infty} ka_n = k \lim_{n\to\infty} a_n$, where k is any real number

Review Video: Limit of a Sequence
Visit mometrix.com/academy and enter code: 847732

Infinite Series

Both arithmetic and geometric sequences have formulas to find the sum of the first n terms in the sequence, assuming you know what the first term is. The sum of all the terms in a sequence is called a series. An **infinite series** is an infinite sum. In other words, it is what you get by adding up all the terms in an infinite sequence: $\sum_{n=1}^{\infty} a_n = a_1 + a_2 + a_3 + \cdots + a_n + \cdots$. This notation can be shortened to $\sum_{n=1}^{\infty} a_n$ or $\sum a_n$.

While we can't add up an infinite list of numbers one at a time, we can still determine the infinite sum. As we add the terms in a series, we can imagine an infinite sequence of partial sums, where the

Copyright © Mometrix Media. You have been licensed one copy of this document for personal use only. Any other reproduction or redistribution is strictly prohibited. All rights reserved.
This content is provided for test preparation purposes only and does not imply an endorsement by Mometrix of any particular political, scientific, or religious point of view.

first partial sum is the first element of the series, the second partial sum is the sum of the first two elements of the series, and the nth partial sum is the sum of the first n elements of the series.

Every infinite sequence of partial sums (infinite series) either converges or diverges. As with the test for convergence in a sequence, finding the limit of the sequence of partial sums will indicate whether it is a converging series or a diverging series. If there exists a real number S such that $\lim_{n\to\infty} S_n = S$, where S_n is the sequence of partial sums, then the series converges. If the limit equals infinity, then the series diverges. If $\lim_{n\to\infty} S_n = S$ and S is a real number, then S is also the convergence value of the series.

To find the sum as n approaches infinity for the sum of two convergent series, find the sum as n approaches infinity for each individual series and add the results.

$$\sum_{n=1}^{\infty}(a_n + b_n) = \sum_{n=1}^{\infty} a_n + \sum_{n=1}^{\infty} b_n$$

The same idea works for subtraction.

$$\sum_{n=1}^{\infty}(a_n - b_n) = \sum_{n=1}^{\infty} a_n - \sum_{n=1}^{\infty} b_n$$

To find the sum as n approaches infinity for the product of a constant (also called a scalar) and a convergent series, find the sum as n approaches infinity for the series and multiply the result by the scalar.

$$\sum_{n=1}^{\infty} ka_n = k\sum_{n=1}^{\infty} a_n$$

Review Video: Infinite Series
Visit mometrix.com/academy and enter code: 271404

The $\mathbf{n^{th}}$ **term test for divergence** means taking the limit of a sequence a_n as n goes to infinity $\left(\lim_{n\to\infty} a_n\right)$ and checking whether the limit is zero. If the limit is not zero, then the series $\sum a_n$ is a diverging series. This test only works to prove divergence, however. If the limit is zero, the test is inconclusive, meaning the series could be either convergent or divergent.

Review Video: Nth Term Test for Divergence
Visit mometrix.com/academy and enter code: 346400

Matrix Basics

A **matrix** (plural: matrices) is a rectangular array of numbers or variables, often called **elements**, which are arranged in columns and rows. A matrix is generally represented by a capital letter, with its elements represented by the corresponding lowercase letter with two subscripts indicating the

Copyright © Mometrix Media. You have been licensed one copy of this document for personal use only. Any other reproduction or redistribution is strictly prohibited. All rights reserved.
This content is provided for test preparation purposes only and does not imply an endorsement by Mometrix of any particular political, scientific, or religious point of view.

row and column of the element. For example, n_{ab} represents the element in row a column b of matrix N.

$$N = \begin{bmatrix} n_{11} & n_{12} & n_{13} \\ n_{21} & n_{22} & n_{23} \end{bmatrix}$$

A matrix can be described in terms of the number of rows and columns it contains in the format $a \times b$, where a is the number of rows and b is the number of columns. The matrix shown above is a 2×3 matrix. Any $a \times b$ matrix where $a = b$ is a square matrix. A **vector** is a matrix that has exactly one column (**column vector**) or exactly one row (**row vector**).

The **main diagonal** of a matrix is the set of elements on the diagonal from the top left to the bottom right of a matrix. Because of the way it is defined, only square matrices will have a main diagonal. For the matrix shown below, the main diagonal consists of the elements $n_{11}, n_{22}, n_{33}, n_{44}$.

$$\begin{bmatrix} n_{11} & n_{12} & n_{13} & n_{14} \\ n_{21} & n_{22} & n_{23} & n_{24} \\ n_{31} & n_{32} & n_{33} & n_{34} \\ n_{41} & n_{42} & n_{43} & n_{44} \end{bmatrix}$$

A 3×4 matrix such as the one shown below would not have a main diagonal because there is no straight line of elements between the top left corner and the bottom right corner that joins the elements.

$$\begin{bmatrix} n_{11} & n_{12} & n_{13} & n_{14} \\ n_{21} & n_{22} & n_{23} & n_{24} \\ n_{31} & n_{32} & n_{33} & n_{34} \end{bmatrix}$$

A **diagonal matrix** is a square matrix that has a zero for every element in the matrix except the elements on the main diagonal. All the elements on the main diagonal must be nonzero numbers.

$$\begin{bmatrix} n_{11} & 0 & 0 & 0 \\ 0 & n_{22} & 0 & 0 \\ 0 & 0 & n_{33} & 0 \\ 0 & 0 & 0 & n_{44} \end{bmatrix}$$

If every element on the main diagonal of a diagonal matrix is equal to one, the matrix is called an **identity matrix**. The identity matrix is often represented by the letter I.

$$I = \begin{bmatrix} 1 & 0 & 0 & 0 \\ 0 & 1 & 0 & 0 \\ 0 & 0 & 1 & 0 \\ 0 & 0 & 0 & 1 \end{bmatrix}$$

A **zero matrix** is a matrix that has zero as the value for every element in the matrix.

$$\begin{bmatrix} 0 & 0 & 0 & 0 \\ 0 & 0 & 0 & 0 \\ 0 & 0 & 0 & 0 \\ 0 & 0 & 0 & 0 \end{bmatrix}$$

The zero matrix is the *identity for matrix addition*. Do not confuse the zero matrix with the identity matrix.

Copyright © Mometrix Media. You have been licensed one copy of this document for personal use only. Any other reproduction or redistribution is strictly prohibited. All rights reserved.
This content is provided for test preparation purposes only and does not imply an endorsement by Mometrix of any particular political, scientific, or religious point of view.

The **negative of a matrix** is also known as the additive inverse of a matrix. If matrix N is the given matrix, then matrix $-N$ is its negative. This means that every element n_{ab} is equal to $-n_{ab}$ in the negative. To find the negative of a given matrix, change the sign of every element in the matrix and keep all elements in their original corresponding positions in the matrix.

If two matrices have the same order and all corresponding elements in the two matrices are the same, then the two matrices are **equal matrices**.

A matrix N may be **transposed** to matrix N^T by changing all rows into columns and changing all columns into rows. The easiest way to accomplish this is to swap the positions of the row and column notations for each element. For example, suppose the element in the second row of the third column of matrix N is $n_{23} = 6$. In the transposed matrix N^T, the transposed element would be $n_{32} = 6$, and it would be placed in the third row of the second column.

$$N = \begin{bmatrix} 1 & 2 & 3 \\ 4 & 5 & 6 \end{bmatrix};\ N^T = \begin{bmatrix} 1 & 4 \\ 2 & 5 \\ 3 & 6 \end{bmatrix}$$

To quickly transpose a matrix by hand, begin with the first column and rewrite a new matrix with those same elements in the same order in the first row. Write the elements from the second column of the original matrix in the second row of the transposed matrix. Continue this process until all columns have been completed. If the original matrix is identical to the transposed matrix, the matrices are symmetric.

Basic Operations with Matrices

There are two categories of basic operations with regard to matrices: operations between a matrix and a scalar, and operations between two matrices.

Scalar Operations

A scalar being added to a matrix is treated as though it were being added to each element of the matrix. The same is true for subtraction, multiplication and division:

$$M + \mathrm{k} = \begin{bmatrix} m_{11} + k & m_{12} + k \\ m_{21} + k & m_{22} + k \end{bmatrix}$$
$$M - k = \begin{bmatrix} m_{11} - k & m_{12} - k \\ m_{21} - k & m_{22} - k \end{bmatrix}$$
$$M \times k = \begin{bmatrix} m_{11} \times k & m_{12} \times k \\ m_{21} \times k & m_{22} \times k \end{bmatrix}$$
$$M \div k = \begin{bmatrix} m_{11} \div k & m_{12} \div k \\ m_{21} \div k & m_{22} \div k \end{bmatrix}$$

Matrix Addition and Subtraction

All four of the basic operations can be used with operations between matrices (although division is usually discarded in favor of multiplication by the inverse), but there are restrictions on the situations in which they can be used. Matrices that meet all the qualifications for a given operation are called **conformable matrices**. However, conformability is specific to the operation; two matrices that are conformable for addition are not necessarily conformable for multiplication.

For two matrices to be conformable for addition or subtraction, they must be of the same dimension; otherwise, the operation is not defined. If matrix M is a 3×2 matrix and matrix N is a

Copyright © Mometrix Media. You have been licensed one copy of this document for personal use only. Any other reproduction or redistribution is strictly prohibited. All rights reserved.
This content is provided for test preparation purposes only and does not imply an endorsement by Mometrix of any particular political, scientific, or religious point of view.

2×3 matrix, the operations $M + N$ and $M - N$ are meaningless. If matrices M and N are the same size, the operation is as simple as adding or subtracting all of the corresponding elements:

$$\begin{bmatrix} m_{11} & m_{12} \\ m_{21} & m_{22} \end{bmatrix} + \begin{bmatrix} n_{11} & n_{12} \\ n_{21} & n_{22} \end{bmatrix} = \begin{bmatrix} m_{11} + n_{11} & m_{12} + n_{12} \\ m_{21} + n_{21} & m_{22} + n_{22} \end{bmatrix}$$

$$\begin{bmatrix} m_{11} & m_{12} \\ m_{21} & m_{22} \end{bmatrix} - \begin{bmatrix} n_{11} & n_{12} \\ n_{21} & n_{22} \end{bmatrix} = \begin{bmatrix} m_{11} - n_{11} & m_{12} - n_{12} \\ m_{21} - n_{21} & m_{22} - n_{22} \end{bmatrix}$$

The result of addition or subtraction is a matrix of the same dimension as the two original matrices involved in the operation.

Matrix Multiplication

The first thing it is necessary to understand about matrix multiplication is that it is not commutative. In scalar multiplication, the operation is commutative, meaning that $a \times b = b \times a$. For matrix multiplication, this is not the case: $A \times B \neq B \times A$. The terminology must be specific when describing matrix multiplication. The operation $A \times B$ can be described as A multiplied (or **post-multiplied**) by B, or B **pre-multiplied** by A.

For two matrices to be conformable for multiplication, they need not be of the same dimension, but specific dimensions must correspond. Taking the example of two matrices M and N to be multiplied $M \times N$, matrix M must have the same number of columns as matrix N has rows. Put another way, if matrix M has the dimensions $a \times b$ and matrix N has the dimensions $c \times d$, b must equal c if the two matrices are to be conformable for this multiplication. The matrix that results from the multiplication will have the dimensions $a \times d$. If a and d are both equal to 1, the product is simply a scalar. Square matrices of the same dimensions are always conformable for multiplication, and their product is always a matrix of the same size.

The simplest type of matrix multiplication is a 1×2 matrix (a row vector) times a 2×1 matrix (a column vector). These will multiply in the following way:

$$\begin{bmatrix} m_{11} & m_{12} \end{bmatrix} \times \begin{bmatrix} n_{11} \\ n_{21} \end{bmatrix} = m_{11}n_{11} + m_{12}n_{21}$$

The two matrices are conformable for multiplication because matrix M has the same number of columns as matrix N has rows. Because the other dimensions are both 1, the result is a scalar. Expanding our matrices to 1×3 and 3×1, the process is the same:

$$\begin{bmatrix} m_{11} & m_{12} & m_{13} \end{bmatrix} \times \begin{bmatrix} n_{11} \\ n_{21} \\ n_{31} \end{bmatrix} = m_{11}n_{11} + m_{12}n_{21} + m_{13}n_{31}$$

Once again, the result is a scalar. This type of basic matrix multiplication is the building block for the multiplication of larger matrices.

To multiply larger matrices, treat each **row from the first matrix** and each **column from the second matrix** as individual vectors and follow the pattern for multiplying vectors. The scalar value found from multiplying the first-row vector by the first column vector is placed in the first row, first column of the new matrix. The scalar value found from multiplying the second-row vector by the first column vector is placed in the second row, first column of the new matrix. Continue this pattern until each row of the first matrix has been multiplied by each column of the second matrix.

Copyright © Mometrix Media. You have been licensed one copy of this document for personal use only. Any other reproduction or redistribution is strictly prohibited. All rights reserved.
This content is provided for test preparation purposes only and does not imply an endorsement by Mometrix of any particular political, scientific, or religious point of view.

Below is an example of the multiplication of a 3×2 matrix and a 2×3 matrix.

$$\begin{bmatrix} m_{11} & m_{12} \\ m_{21} & m_{22} \\ m_{31} & m_{32} \end{bmatrix} \times \begin{bmatrix} n_{11} & n_{12} & n_{13} \\ n_{21} & n_{22} & n_{23} \end{bmatrix} = \begin{bmatrix} m_{11}n_{11} + m_{12}n_{21} & m_{11}n_{12} + m_{12}n_{22} & m_{11}n_{13} + m_{12}n_{23} \\ m_{21}n_{11} + m_{22}n_{21} & m_{21}n_{12} + m_{22}n_{22} & m_{21}n_{13} + m_{22}n_{23} \\ m_{31}n_{11} + m_{32}n_{21} & m_{31}n_{12} + m_{32}n_{22} & m_{31}n_{13} + m_{32}n_{23} \end{bmatrix}$$

This process starts by taking the first column of the second matrix and running it through each row of the first matrix. Removing all but the first M row and first N column, we would see only the following:

$$[m_{11} \;\; m_{12}] \times \begin{bmatrix} n_{11} \\ n_{21} \end{bmatrix} = m_{11}n_{11} + m_{12}n_{21}$$

The first product would then be $m_{11}n_{11} + m_{12}n_{21}$. This process will be continued for each column of the N matrix to find the first full row of the product matrix, as shown below.

$$[m_{11}n_{11} + m_{12}n_{21} \quad m_{11}n_{12} + m_{12}n_{22} \quad m_{11}n_{13} + m_{12}n_{23}]$$

After completing the first row, the next step would be to simply move to the second row of the M matrix and repeat the process until all of the rows have been finished. The result is a 3×3 matrix.

$$\begin{bmatrix} m_{11} & m_{12} \\ m_{21} & m_{22} \\ m_{31} & m_{32} \end{bmatrix} \times \begin{bmatrix} n_{11} & n_{12} & n_{13} \\ n_{21} & n_{22} & n_{23} \end{bmatrix} = \begin{bmatrix} m_{11}n_{11} + m_{12}n_{21} & m_{11}n_{12} + m_{12}n_{22} & m_{11}n_{13} + m_{12}n_{23} \\ m_{21}n_{11} + m_{22}n_{21} & m_{21}n_{12} + m_{22}n_{22} & m_{21}n_{13} + m_{22}n_{23} \\ m_{31}n_{11} + m_{32}n_{21} & m_{31}n_{12} + m_{32}n_{22} & m_{31}n_{13} + m_{32}n_{23} \end{bmatrix}$$

If the operation were done in reverse $(N \times M)$, the result would be a 2×2 matrix.

$$\begin{bmatrix} n_{11} & n_{12} & n_{13} \\ n_{21} & n_{22} & n_{23} \end{bmatrix} \times \begin{bmatrix} m_{11} & m_{12} \\ m_{21} & m_{22} \\ m_{31} & m_{32} \end{bmatrix} = \begin{bmatrix} m_{11}n_{11} + m_{21}n_{12} + m_{31}n_{13} & m_{12}n_{11} + m_{22}n_{12} + m_{32}n_{13} \\ m_{11}n_{21} + m_{21}n_{22} + m_{31}n_{23} & m_{12}n_{21} + m_{22}n_{22} + m_{32}n_{23} \end{bmatrix}$$

Review Video: Matrices: The Basics
Visit mometrix.com/academy and enter code: 516658

Advanced Matrices

Elementary Row Operations

Elementary row operations include multiplying a row by a non-zero scalar, adding scalar multiples of two rows, and switching rows. These operations can be done using matrix multiplication with specialized transformation matrices. Elementary row operations are left multiplied, $A' = MA$.To do the same type of operations on the columns, then the transformation matrix is right multiplied, $A' = AM$. **Row switching** is achieved by swapping the corresponding rows in the identity matrix. For example, consider switching row 2 and row 3 in a 3×3 matrix:

$$M_{R_2 \leftrightarrow R_3} = \begin{bmatrix} 1 & 0 & 0 \\ 0 & 0 & 1 \\ 0 & 1 & 0 \end{bmatrix}$$

Copyright © Mometrix Media. You have been licensed one copy of this document for personal use only. Any other reproduction or redistribution is strictly prohibited. All rights reserved.
This content is provided for test preparation purposes only and does not imply an endorsement by Mometrix of any particular political, scientific, or religious point of view.

The transformation matrix for **row multiplication** is also based on the identity matrix with the scalar multiplication factor in place of the 1 in the corresponding row. Multiplying row 1 by –4 in a 3 × 3 matrix:

$$M_{-4R_1 \to R_1} = \begin{bmatrix} -4 & 0 & 0 \\ 0 & 1 & 0 \\ 0 & 0 & 1 \end{bmatrix}$$

The transformation matrix for **row addition** consists of the identity matrix with a 1 in the element corresponding to the two rows being added in the column where you want the result to go. Adding row 2 to row 1 in a 3 × 3 matrix:

$$M_{R_1+R_2 \to R_1} = \begin{bmatrix} 1 & 1 & 0 \\ 0 & 1 & 0 \\ 0 & 0 & 1 \end{bmatrix}$$

Review Video: Matrices: Elementary Row Operations
Visit mometrix.com/academy and enter code: 493170

Determinants and Inverses of Matrices

The **determinant** of a matrix is a scalar value that is calculated by taking into account all the elements of a square matrix. A determinant only exists for square matrices. Finding the determinant of a 2 × 2 matrix is as simple as remembering a simple equation. For a 2 × 2 matrix $M = \begin{bmatrix} m_{11} & m_{12} \\ m_{21} & m_{22} \end{bmatrix}$, the determinant is obtained by the equation $|M| = m_{11}m_{22} - m_{12}m_{21}$. Anything larger than 2 × 2 requires multiple steps. Take matrix $N = \begin{bmatrix} a & b & c \\ d & e & f \\ g & h & j \end{bmatrix}$. The determinant of N is calculated as $|N| = a\begin{vmatrix} e & f \\ h & j \end{vmatrix} - b\begin{vmatrix} d & f \\ g & j \end{vmatrix} + c\begin{vmatrix} d & e \\ g & h \end{vmatrix}$ or $|N| = a(ej - fh) - b(dj - fg) + c(dh - eg)$.

There is a shortcut for 3 × 3 matrices: add the products of each unique set of elements diagonally left-to-right and subtract the products of each unique set of elements diagonally right-to-left. In matrix N, the left-to-right diagonal elements are (a, e, j), (b, f, g), and (c, d, h). The right-to-left diagonal elements are (a, f, h), (b, d, j), and (c, e, g). $\det(N) = aej + bfg + cdh - afh - bdj - ceg$.

Calculating the determinants of matrices larger than 3 × 3 is rarely, if ever, done by hand.

The **inverse** of a matrix M is the matrix that, when multiplied by matrix M, yields a product that is the identity matrix. Multiplication of matrices will be explained in greater detail shortly. Not all matrices have inverses. Only a square matrix whose determinant is not zero has an inverse. If a matrix has an inverse, that inverse is unique to that matrix. For any matrix M that has an inverse, the inverse is represented by the symbol M^{-1}. To calculate the inverse of a 2 × 2 square matrix, use the following pattern:

$$M = \begin{bmatrix} m_{11} & m_{12} \\ m_{21} & m_{22} \end{bmatrix}; \; M^{-1} = \begin{bmatrix} \frac{m_{22}}{|M|} & \frac{-m_{12}}{|M|} \\ \frac{-m_{21}}{|M|} & \frac{m_{11}}{|M|} \end{bmatrix}$$

Another way to find the inverse of a matrix by hand is use an augmented matrix and elementary row operations. An **augmented matrix** is formed by appending the entries from one matrix onto

Copyright © Mometrix Media. You have been licensed one copy of this document for personal use only. Any other reproduction or redistribution is strictly prohibited. All rights reserved.
This content is provided for test preparation purposes only and does not imply an endorsement by Mometrix of any particular political, scientific, or religious point of view.

the end of another. For example, given a 2×2 invertible matrix $N = \begin{bmatrix} a & b \\ c & d \end{bmatrix}$, you can find the inverse N^{-1} by creating an augmented matrix by appending a 2×2 identity matrix: $\left[\begin{matrix} a & b \\ c & d \end{matrix}\middle|\begin{matrix} 1 & 0 \\ 0 & 1 \end{matrix}\right]$. To find the inverse of the original 2×2 matrix, perform elementary row operations to convert the original matrix on the left to an identity matrix: $\left[\begin{matrix} 1 & 0 \\ 0 & 1 \end{matrix}\middle|\begin{matrix} e & f \\ g & h \end{matrix}\right]$. For instance, the first step might be to multiply the second row by $\frac{b}{d}$ and then subtract it from the first row to make its second column a zero. The end result is that the 2×2 section on the right will become the inverse of the original matrix: $N^{-1} = \begin{bmatrix} e & f \\ g & h \end{bmatrix}$.

Review Video: Transposition, Determinants, and Augmentation of Matrices
Visit mometrix.com/academy and enter code: 521365

Reduced Row-Echelon Forms

When a system of equations has a solution, finding the transformation of the augmented matrix will result in one of three reduced row-echelon forms. Only one of these forms will give a unique solution to the system of equations, however. The following examples show the solutions indicated by particular results:

$\begin{bmatrix} 1 & 0 & 0 & x_0 \\ 0 & 1 & 0 & y_0 \\ 0 & 0 & 1 & z_0 \end{bmatrix}$ gives the unique solution $x = x_0;\ y = y_0; z = z_0$

$\begin{bmatrix} 1 & 0 & k_1 & x_0 \\ 0 & 1 & k_2 & y_0 \\ 0 & 0 & 0 & 0 \end{bmatrix}$ gives a non-unique solution $x = x_0 - k_1 z; y = y_0 - k_2 z$

$\begin{bmatrix} 1 & j_1 & k_1 & x_0 \\ 0 & 0 & 0 & 0 \\ 0 & 0 & 0 & 0 \end{bmatrix}$ gives a non-unique solution $x = x_0 - j_1 y - k_1 z$

Geometric Transformations

The four *geometric transformations* are **translations**, **reflections**, **rotations**, and **dilations**. When geometric transformations are expressed as matrices, the process of performing the transformations is simplified. For calculations of the geometric transformations of a planar figure, make a $2 \times n$ matrix, where n is the number of vertices in the planar figure. Each column represents the rectangular coordinates of one vertex of the figure, with the top row containing the values of the x-coordinates and the bottom row containing the values of the y-coordinates. For example, given a planar triangular figure with coordinates (x_1, y_1), (x_2, y_2), and (x_3, y_3), the corresponding matrix is $\begin{bmatrix} x_1 & x_2 & x_3 \\ y_1 & y_2 & y_3 \end{bmatrix}$. You can then perform the necessary transformations on this matrix to determine the coordinates of the resulting figure.

Review Video: Matrices: Geometric Transformations
Visit mometrix.com/academy and enter code: 612781

Copyright © Mometrix Media. You have been licensed one copy of this document for personal use only. Any other reproduction or redistribution is strictly prohibited. All rights reserved.
This content is provided for test preparation purposes only and does not imply an endorsement by Mometrix of any particular political, scientific, or religious point of view.

Translation

A **translation** moves a figure along the x-axis, the y-axis, or both axes without changing the size or shape of the figure. To calculate the new coordinates of a planar figure following a translation, set up a matrix of the coordinates and a matrix of the translation values and add the two matrices.

$$\begin{bmatrix} h & h & h \\ v & v & v \end{bmatrix} + \begin{bmatrix} x_1 & x_2 & x_3 \\ y_1 & y_2 & y_3 \end{bmatrix} = \begin{bmatrix} h+x_1 & h+x_2 & h+x_3 \\ v+y_1 & v+y_2 & v+y_3 \end{bmatrix}$$

where h is the number of units the figure is moved along the x-axis (horizontally) and v is the number of units the figure is moved along the y-axis (vertically).

Reflection

To find the **reflection** of a planar figure over the x-axis, set up a matrix of the coordinates of the vertices and pre-multiply the matrix by the 2×2 matrix $\begin{bmatrix} 1 & 0 \\ 0 & -1 \end{bmatrix}$ so that $\begin{bmatrix} 1 & 0 \\ 0 & -1 \end{bmatrix}\begin{bmatrix} x_1 & x_2 & x_3 \\ y_1 & y_2 & y_3 \end{bmatrix} = \begin{bmatrix} x_1 & x_2 & x_3 \\ -y_1 & -y_2 & -y_3 \end{bmatrix}$. To find the reflection of a planar figure over the y-axis, set up a matrix of the coordinates of the vertices and pre-multiply the matrix by the 2×2 matrix $\begin{bmatrix} -1 & 0 \\ 0 & 1 \end{bmatrix}$ so that $\begin{bmatrix} -1 & 0 \\ 0 & 1 \end{bmatrix}\begin{bmatrix} x_1 & x_2 & x_3 \\ y_1 & y_2 & y_3 \end{bmatrix} = \begin{bmatrix} -x_1 & -x_2 & -x_3 \\ y_1 & y_2 & y_3 \end{bmatrix}$. To find the reflection of a planar figure over the line $y = x$, set up a matrix of the coordinates of the vertices and pre-multiply the matrix by the 2×2 matrix $\begin{bmatrix} 0 & 1 \\ 1 & 0 \end{bmatrix}$ so that $\begin{bmatrix} 0 & 1 \\ 1 & 0 \end{bmatrix}\begin{bmatrix} x_1 & x_2 & x_3 \\ y_1 & y_2 & y_3 \end{bmatrix} = \begin{bmatrix} y_1 & y_2 & y_3 \\ x_1 & x_2 & x_3 \end{bmatrix}$. Remember that the order of multiplication is important when multiplying matrices. The commutative property does not apply.

Rotation

To find the coordinates of the figure formed by rotating a planar figure about the origin θ degrees in a counterclockwise direction, set up a matrix of the coordinates of the vertices and pre-multiply the matrix by the 2×2 matrix $\begin{bmatrix} \cos\theta & \sin\theta \\ -\sin\theta & \cos\theta \end{bmatrix}$. For example, if you want to rotate a figure 90° clockwise around the origin, you would have to convert the degree measure to 270° counterclockwise and solve the 2×2 matrix you have set as the pre-multiplier: $\begin{bmatrix} \cos 270° & \sin 270° \\ -\sin 270° & \cos 270° \end{bmatrix} = \begin{bmatrix} 0 & -1 \\ 1 & 0 \end{bmatrix}$. Use this as the pre-multiplier for the matrix $\begin{bmatrix} x_1 & x_2 & x_3 \\ y_1 & y_2 & y_3 \end{bmatrix}$ and solve to find the new coordinates.

Dilation

To find the **dilation** of a planar figure by a scale factor of k, set up a matrix of the coordinates of the vertices of the planar figure and pre-multiply the matrix by the 2×2 matrix $\begin{bmatrix} k & 0 \\ 0 & k \end{bmatrix}$ so that $\begin{bmatrix} k & 0 \\ 0 & k \end{bmatrix}\begin{bmatrix} x_1 & x_2 & x_3 \\ y_1 & y_2 & y_3 \end{bmatrix} = \begin{bmatrix} kx_1 & kx_2 & kx_3 \\ ky_1 & ky_2 & ky_3 \end{bmatrix}$. This is effectively the same as multiplying the matrix by the scalar k, but the matrix equation would still be necessary if the figure were being dilated by different factors in vertical and horizontal directions. The scale factor k will be greater than 1 if the figure is being enlarged, and between 0 and 1 if the figure is being shrunk. Again, remember that when multiplying matrices, the order of the matrices is important. The commutative property does not apply, and the matrix with the coordinates of the figure must be the second matrix.

Copyright © Mometrix Media. You have been licensed one copy of this document for personal use only. Any other reproduction or redistribution is strictly prohibited. All rights reserved.
This content is provided for test preparation purposes only and does not imply an endorsement by Mometrix of any particular political, scientific, or religious point of view.

Matrix Systems

Solving Systems of Equations

Matrices can be used to represent the coefficients of a system of linear equations and can be very useful in solving those systems. Take for instance three equations with three variables where all a, b, c, and d are known constants:

$$\begin{aligned} a_1x + b_1y + c_1z &= d_1 \\ a_2x + b_2y + c_2z &= d_2 \\ a_3x + b_3y + c_3z &= d_3 \end{aligned}$$

To solve this system, define three matrices:

$$A = \begin{bmatrix} a_1 & b_1 & c_1 \\ a_2 & b_2 & c_2 \\ a_3 & b_3 & c_3 \end{bmatrix}; \; D = \begin{bmatrix} d_1 \\ d_2 \\ d_3 \end{bmatrix}; \; X = \begin{bmatrix} x \\ y \\ z \end{bmatrix}$$

The three equations in our system can be fully represented by a single matrix equation:

$$AX = D$$

We know that the identity matrix times X is equal to X, and we know that any matrix multiplied by its inverse is equal to the identity matrix.

$$A^{-1}AX = IX = X; \text{thus } X = A^{-1}D$$

Our goal then is to find the inverse of A, or A^{-1}. Once we have that, we can pre-multiply matrix D by A^{-1} (post-multiplying here is an undefined operation) to find matrix X.

Systems of equations can also be solved using the transformation of an augmented matrix in a process similar to that for finding a matrix inverse. Begin by arranging each equation of the system in the following format:

$$\begin{aligned} a_1x + b_1y + c_1z &= d_1 \\ a_2x + b_2y + c_2z &= d_2 \\ a_3x + b_3y + c_3z &= d_3 \end{aligned}$$

Define matrices A and D and combine them into augmented matrix A_a:

$$A = \begin{bmatrix} a_1 & b_1 & c_1 \\ a_2 & b_2 & c_2 \\ a_3 & b_3 & c_3 \end{bmatrix}; \; D = \begin{bmatrix} d_1 \\ d_2 \\ d_3 \end{bmatrix}; A_a = \begin{bmatrix} a_1 & b_1 & c_1 & d_1 \\ a_2 & b_2 & c_2 & d_2 \\ a_3 & b_3 & c_3 & d_3 \end{bmatrix}$$

To solve the augmented matrix and the system of equations, use elementary row operations to form an identity matrix in the first 3×3 section. When this is complete, the values in the last column are the solutions to the system of equations:

$$\begin{bmatrix} 1 & 0 & 0 & x \\ 0 & 1 & 0 & y \\ 0 & 0 & 1 & z \end{bmatrix}$$

Copyright © Mometrix Media. You have been licensed one copy of this document for personal use only. Any other reproduction or redistribution is strictly prohibited. All rights reserved.
This content is provided for test preparation purposes only and does not imply an endorsement by Mometrix of any particular political, scientific, or religious point of view.

If an identity matrix is not possible, the system of equations has no unique solution. Sometimes only a partial solution will be possible. The following are partial solutions you may find:

$\begin{bmatrix} 1 & 0 & k_1 & x_0 \\ 0 & 1 & k_2 & y_0 \\ 0 & 0 & 0 & 0 \end{bmatrix}$ gives the non-unique solution $x = x_0 - k_1 z;\ y = y_0 - k_2 z$

$\begin{bmatrix} 1 & j_1 & k_1 & x_0 \\ 0 & 0 & 0 & 0 \\ 0 & 0 & 0 & 0 \end{bmatrix}$ gives the non-unique solution $x = x_0 - j_1 y - k_1 z$

This process can be used to solve systems of equations with any number of variables, but three is the upper limit for practical purposes. Anything more ought to be done with a graphing calculator.

Review Video: Matrices: Data Systems
Visit mometrix.com/academy and enter code: 579763

Points, Lines, and Planes

Points and Lines

A **point** is a fixed location in space, has no size or dimensions, and is commonly represented by a dot. A **line** is a set of points that extends infinitely in two opposite directions. It has length, but no width or depth. A line can be defined by any two distinct points that it contains. A **line segment** is a portion of a line that has definite endpoints. A **ray** is a portion of a line that extends from a single point on that line in one direction along the line. It has a definite beginning, but no ending.

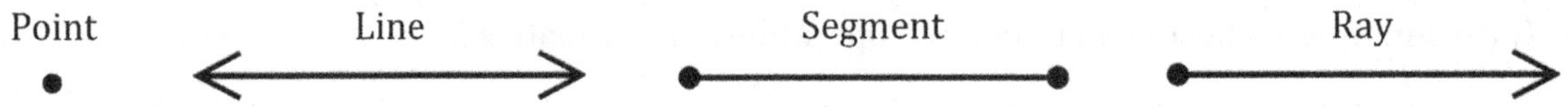

Interactions Between Lines

Intersecting lines are lines that have exactly one point in common. **Concurrent lines** are multiple lines that intersect at a single point. **Perpendicular lines** are lines that intersect at right angles. They are represented by the symbol ⊥. The shortest distance from a line to a point not on the line is a perpendicular segment from the point to the line. **Parallel lines** are lines in the same plane that have no points in common and never meet. It is possible for lines to be in different planes, have no points in common, and never meet, but they are not parallel because they are in different planes.

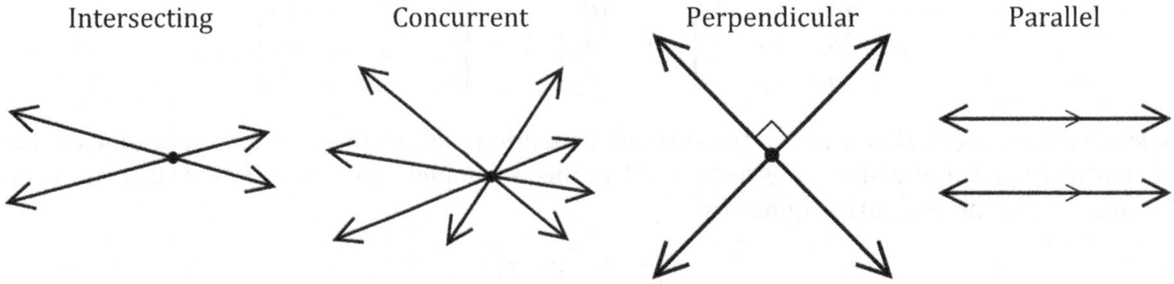

Review Video: Parallel and Perpendicular Lines
Visit mometrix.com/academy and enter code: 815923

Copyright © Mometrix Media. You have been licensed one copy of this document for personal use only. Any other reproduction or redistribution is strictly prohibited. All rights reserved.
This content is provided for test preparation purposes only and does not imply an endorsement by Mometrix of any particular political, scientific, or religious point of view.

A **transversal** is a line that intersects at least two other lines, which may or may not be parallel to one another. A transversal that intersects parallel lines is a common occurrence in geometry. A **bisector** is a line or line segment that divides another line segment into two equal lengths. A **perpendicular bisector** of a line segment is composed of points that are equidistant from the endpoints of the segment it is dividing.

Transversal

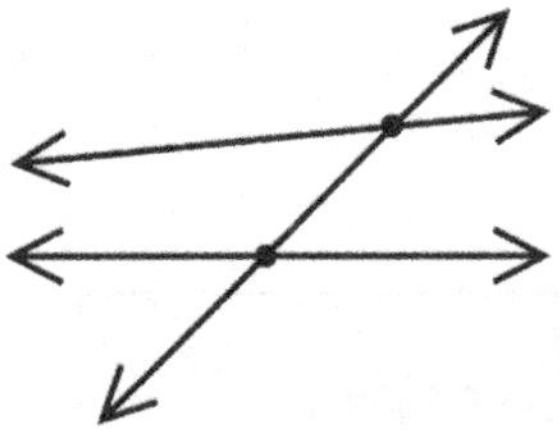

Bisector

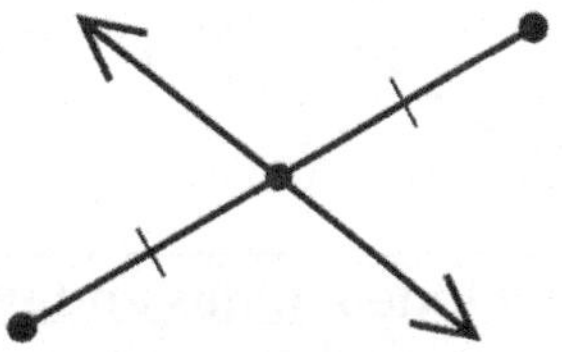

Perpendicular bisector

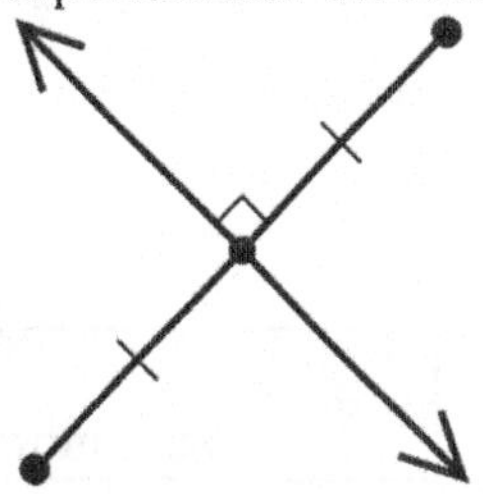

The **projection of a point on a line** is the point at which a perpendicular line drawn from the given point to the given line intersects the line. This is also the shortest distance from the given point to the line. The **projection of a segment on a line** is a segment whose endpoints are the points formed when perpendicular lines are drawn from the endpoints of the given segment to the given line. This is similar to the length a diagonal line appears to be when viewed from above.

Projection of a point on a line

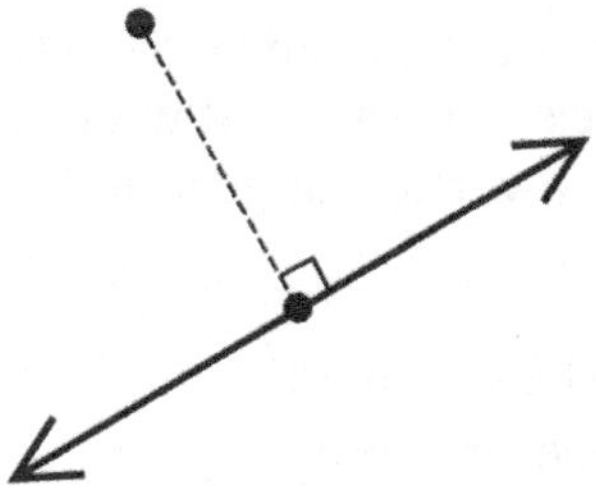

Projection of a segment on a line

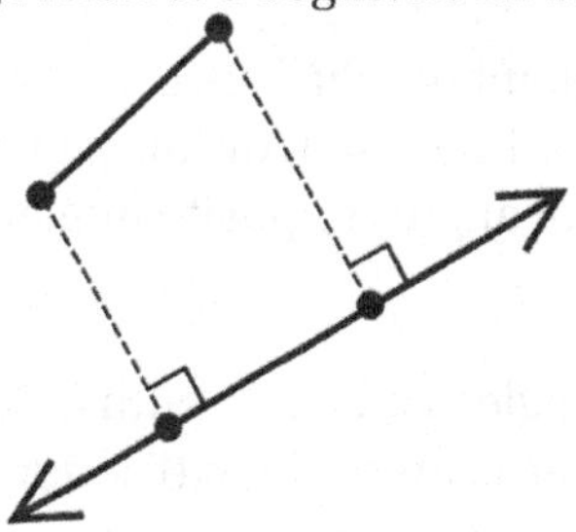

PLANES

A **plane** is a two-dimensional flat surface defined by three non-collinear points. A plane extends an infinite distance in all directions in those two dimensions. It contains an infinite number of points, parallel lines and segments, intersecting lines and segments, as well as parallel or intersecting rays. A plane will never contain a three-dimensional figure or skew lines, which are lines that don't intersect and are not parallel. Two given planes are either parallel or they intersect at a line. A plane

Copyright © Mometrix Media. You have been licensed one copy of this document for personal use only. Any other reproduction or redistribution is strictly prohibited. All rights reserved.
This content is provided for test preparation purposes only and does not imply an endorsement by Mometrix of any particular political, scientific, or religious point of view.

may intersect a circular conic surface to form **conic sections**, such as a parabola, hyperbola, circle or ellipse.

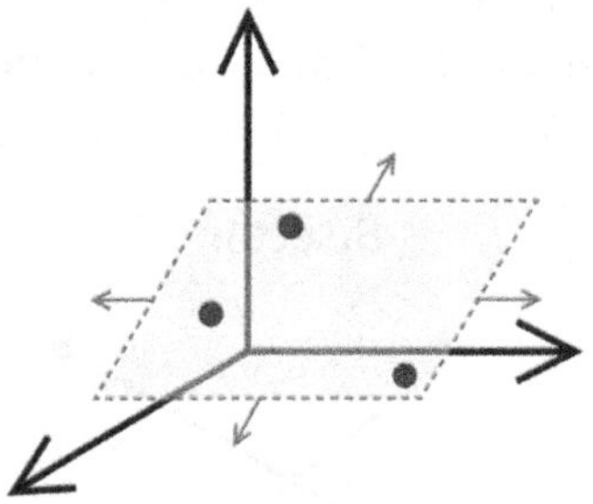

Review Video: Lines and Planes
Visit mometrix.com/academy and enter code: 554267

Angles

ANGLES AND VERTICES

An **angle** is formed when two lines or line segments meet at a common point. It may be a common starting point for a pair of segments or rays, or it may be the intersection of lines. Angles are represented by the symbol ∠.

The **vertex** is the point at which two segments or rays meet to form an angle. If the angle is formed by intersecting rays, lines, and/or line segments, the vertex is the point at which four angles are formed. The pairs of angles opposite one another are called vertical angles, and their measures are equal.

- An **acute** angle is an angle with a degree measure less than 90°.
- A **right** angle is an angle with a degree measure of exactly 90°.
- An **obtuse** angle is an angle with a degree measure greater than 90° but less than 180°.
- A **straight angle** is an angle with a degree measure of exactly 180°. This is also a semicircle.
- A **reflex angle** is an angle with a degree measure greater than 180° but less than 360°.
- A **full angle** is an angle with a degree measure of exactly 360°. This is also a circle.

Review Video: Angles
Visit mometrix.com/academy and enter code: 264624

RELATIONSHIPS BETWEEN ANGLES

Two angles whose sum is exactly 90° are said to be **complementary**. The two angles may or may not be adjacent. In a right triangle, the two acute angles are complementary.

Two angles whose sum is exactly 180° are said to be **supplementary**. The two angles may or may not be adjacent. Two intersecting lines always form two pairs of supplementary angles. Adjacent supplementary angles will always form a straight line.

Copyright © Mometrix Media. You have been licensed one copy of this document for personal use only. Any other reproduction or redistribution is strictly prohibited. All rights reserved.
This content is provided for test preparation purposes only and does not imply an endorsement by Mometrix of any particular political, scientific, or religious point of view.

Two angles that have the same vertex and share a side are said to be **adjacent**. Vertical angles are not adjacent because they share a vertex but no common side.

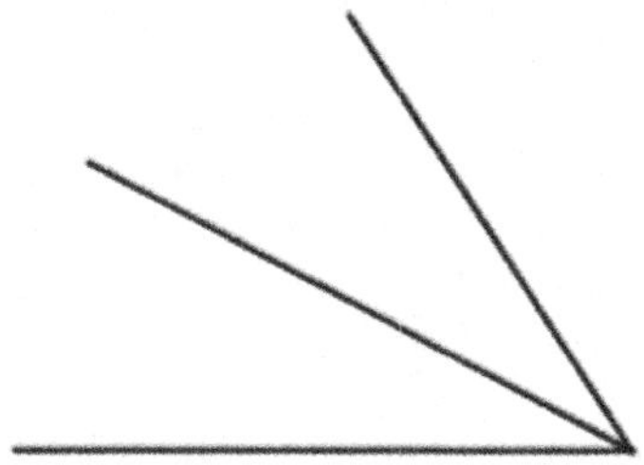

Adjacent
Share vertex and side

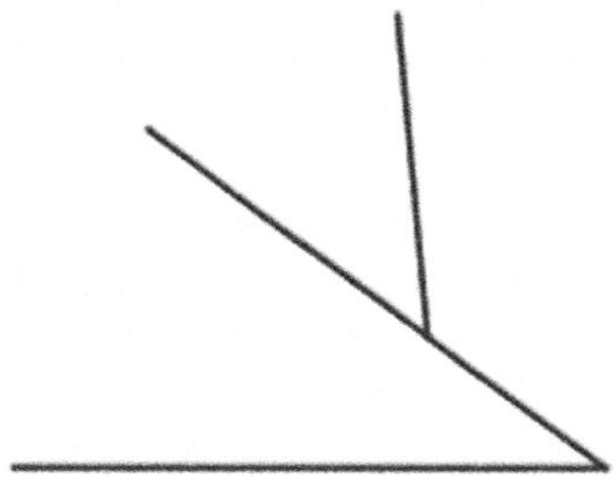

Not adjacent
Share part of a side, but not vertex

When two parallel lines are cut by a transversal, the angles that are between the two parallel lines are **interior angles**. In the diagram below, angles 3, 4, 5, and 6 are interior angles.

When two parallel lines are cut by a transversal, the angles that are outside the parallel lines are **exterior angles**. In the diagram below, angles 1, 2, 7, and 8 are exterior angles.

When two parallel lines are cut by a transversal, the angles that are in the same position relative to the transversal and a parallel line are **corresponding angles**. The diagram below has four pairs of corresponding angles: angles 1 and 5, angles 2 and 6, angles 3 and 7, and angles 4 and 8. Corresponding angles formed by parallel lines are congruent.

When two parallel lines are cut by a transversal, the two interior angles that are on opposite sides of the transversal are called **alternate interior angles**. In the diagram below, there are two pairs of alternate interior angles: angles 3 and 6, and angles 4 and 5. Alternate interior angles formed by parallel lines are congruent.

When two parallel lines are cut by a transversal, the two exterior angles that are on opposite sides of the transversal are called **alternate exterior angles**.

In the diagram below, there are two pairs of alternate exterior angles: angles 1 and 8, and angles 2 and 7. Alternate exterior angles formed by parallel lines are congruent.

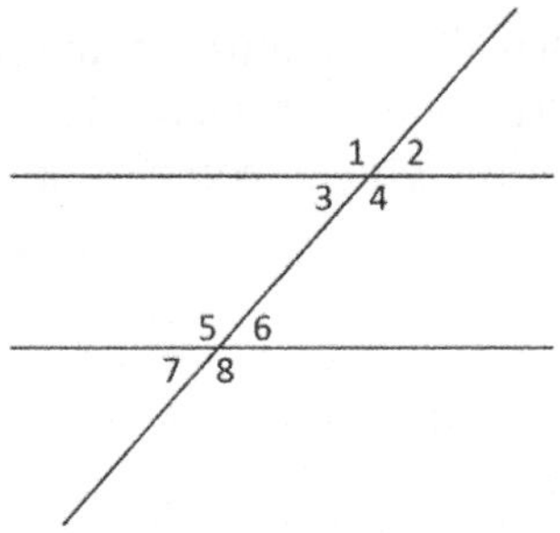

When two lines intersect, four angles are formed. The non-adjacent angles at this vertex are called vertical angles. Vertical angles are congruent. In the diagram, $\angle ABD \cong \angle CBE$ and $\angle ABC \cong \angle DBE$.

Copyright © Mometrix Media. You have been licensed one copy of this document for personal use only. Any other reproduction or redistribution is strictly prohibited. All rights reserved.
This content is provided for test preparation purposes only and does not imply an endorsement by Mometrix of any particular political, scientific, or religious point of view.

The other pairs of angles, ($\angle ABC$, $\angle CBE$) and ($\angle ABD$, $\angle DBE$), are supplementary, meaning the pairs sum to 180°.

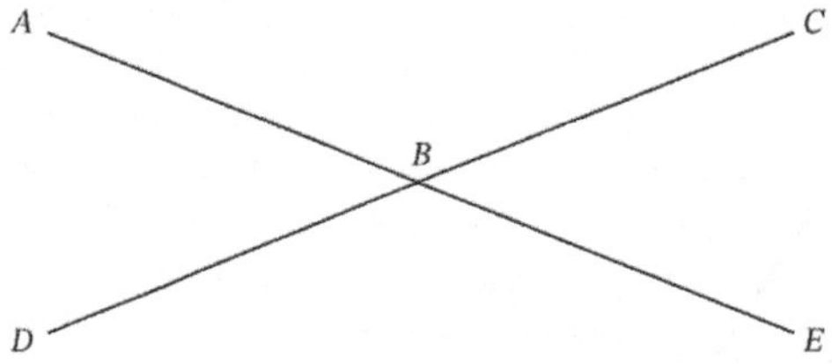

Transformations

Rotation

A **rotation** is a transformation that turns a figure around a point called the **center of rotation**, which can lie anywhere in the plane. If a line is drawn from a point on a figure to the center of rotation, and another line is drawn from the center to the rotated image of that point, the angle between the two lines is the **angle of rotation**. The vertex of the angle of rotation is the center of rotation.

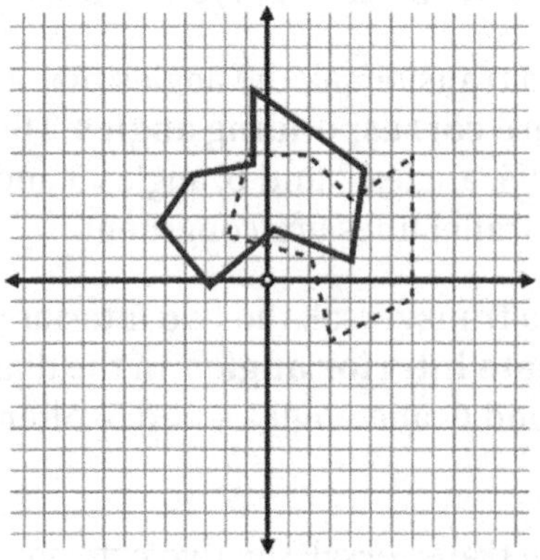

Review Video: Rotation
Visit mometrix.com/academy and enter code: 602600

Translation and Dilation

A **translation** is a transformation which slides a figure from one position in the plane to another position in the plane. The original figure and the translated figure have the same size, shape, and orientation. A **dilation** is a transformation which proportionally stretches or shrinks a figure by a

Copyright © Mometrix Media. You have been licensed one copy of this document for personal use only. Any other reproduction or redistribution is strictly prohibited. All rights reserved.
This content is provided for test preparation purposes only and does not imply an endorsement by Mometrix of any particular political, scientific, or religious point of view.

scale factor. The dilated image is the same shape and orientation as the original image but a different size. A polygon and its dilated image are similar.

Translation

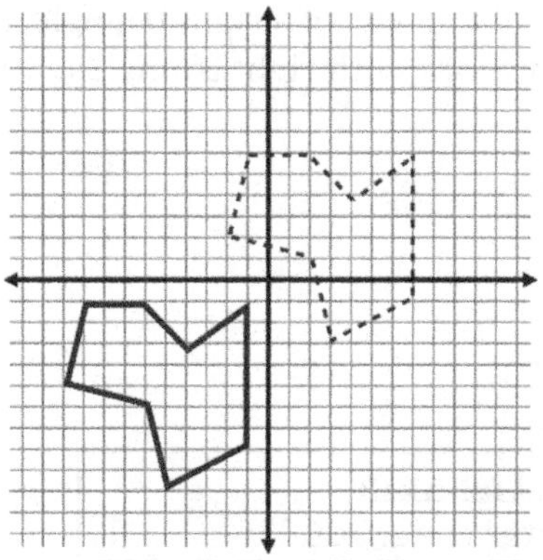

Dilation

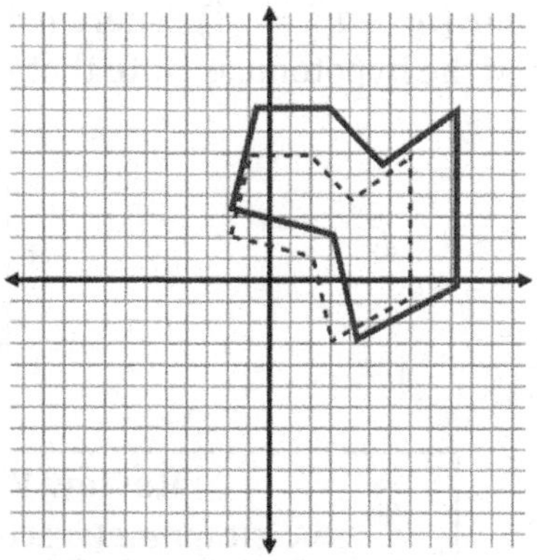

Review Video: Translation
Visit mometrix.com/academy and enter code: 718628

Review Video: Dilation
Visit mometrix.com/academy and enter code: 471630

A **reflection of a figure over a line** (a "flip") creates a congruent image that is the same distance from the line as the original figure but on the opposite side. The **line of reflection** is the perpendicular bisector of any line segment drawn from a point on the original figure to its reflected image (unless the point and its reflected image happen to be the same point, which happens when a figure is reflected over one of its own sides). A **reflection of a figure over a point** (an inversion) in two dimensions is the same as the rotation of the figure 180° about that point. The image of the figure is congruent to the original figure. The **point of reflection** is the midpoint of a line segment which connects a point in the figure to its image (unless the point and its reflected image happen to be the same point, which happens when a figure is reflected in one of its own points).

Reflection of a figure over a line

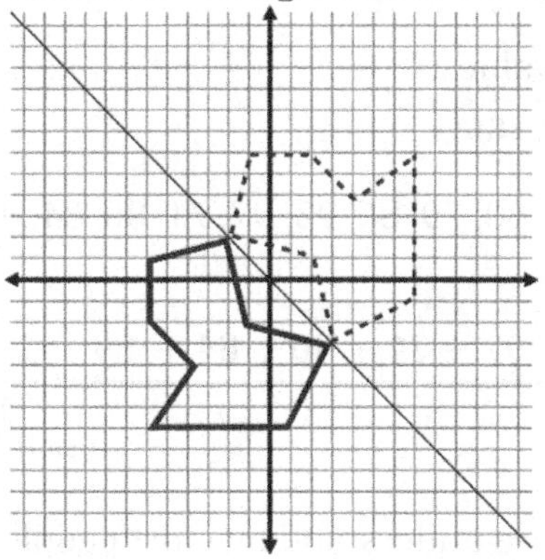

Reflection of a figure over a point

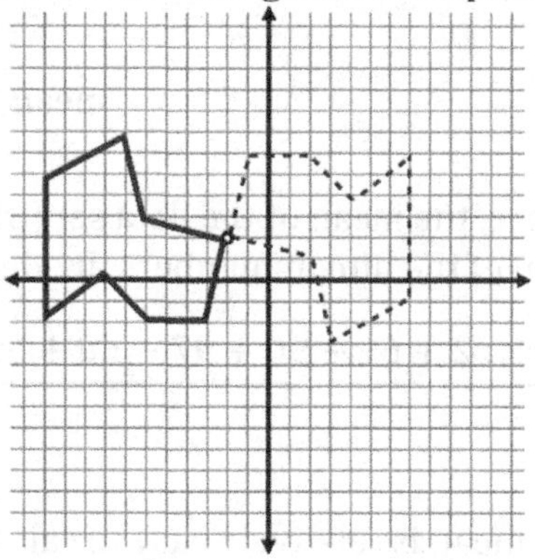

Review Video: Reflection
Visit mometrix.com/academy and enter code: 955068

Copyright © Mometrix Media. You have been licensed one copy of this document for personal use only. Any other reproduction or redistribution is strictly prohibited. All rights reserved.
This content is provided for test preparation purposes only and does not imply an endorsement by Mometrix of any particular political, scientific, or religious point of view.

Pythagorean Theorem

The side of a triangle opposite the right angle is called the **hypotenuse**. The other two sides are called the legs. The Pythagorean theorem states a relationship among the legs and hypotenuse of a right triangle: $(a^2 + b^2 = c^2)$, where a and b are the lengths of the legs of a right triangle, and c is the length of the hypotenuse. Note that this formula will only work with right triangles.

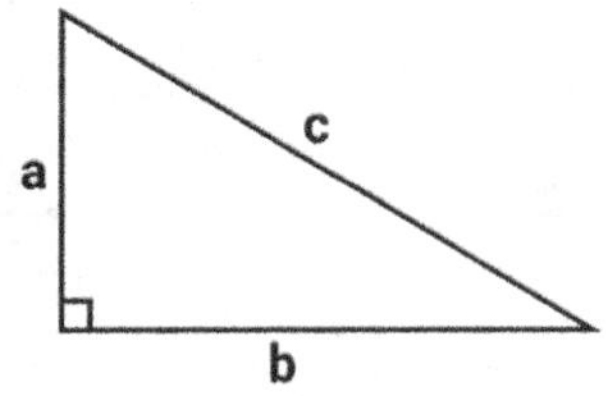

Review Video: Pythagorean Theorem
Visit mometrix.com/academy and enter code: 906576

Measures of Central Tendency

A **measure of central tendency** is a statistical value that gives a reasonable estimate for the center of a group of data. There are several different ways of describing the measure of central tendency. Each one has a unique way it is calculated, and each one gives a slightly different perspective on the data set. Whenever you give a measure of central tendency, always make sure the units are the same. If the data has different units, such as hours, minutes, and seconds, convert all the data to the same unit, and use the same unit in the measure of central tendency. If no units are given in the data, do not give units for the measure of central tendency.

Mean

The **statistical mean** of a group of data is the same as the arithmetic average of that group. To find the mean of a set of data, first convert each value to the same units, if necessary. Then find the sum of all the values, and count the total number of data values, making sure you take into consideration each individual value. If a value appears more than once, count it more than once. Divide the sum of the values by the total number of values and apply the units, if any. Note that the mean does not have to be one of the data values in the set, and may not divide evenly.

$$\text{mean} = \frac{\text{sum of the data values}}{\text{quantity of data values}}$$

For instance, the mean of the data set {88, 72, 61, 90, 97, 68, 88, 79, 86, 93, 97, 71, 80, 84, 89} would be the sum of the fifteen numbers divided by 15:

$$\frac{88 + 72 + 61 + 90 + 97 + 68 + 88 + 79 + 86 + 93 + 97 + 71 + 80 + 84 + 89}{15} = \frac{1242}{15} = 82.8$$

While the mean is relatively easy to calculate and averages are understood by most people, the mean can be very misleading if it is used as the sole measure of central tendency. If the data set has outliers (data values that are unusually high or unusually low compared to the rest of the data values), the mean can be very distorted, especially if the data set has a small number of values. If unusually high values are countered with unusually low values, the mean is not affected as much. For example, if five of twenty students in a class get a 100 on a test, but the other 15 students have an average of 60 on the same test, the class average would appear as 70. Whenever the mean is

Copyright © Mometrix Media. You have been licensed one copy of this document for personal use only. Any other reproduction or redistribution is strictly prohibited. All rights reserved.
This content is provided for test preparation purposes only and does not imply an endorsement by Mometrix of any particular political, scientific, or religious point of view.

skewed by outliers, it is always a good idea to include the median as an alternate measure of central tendency.

A **weighted mean**, or weighted average, is a mean that uses "weighted" values. The formula is weighted mean $= \frac{w_1x_1+w_2x_2+w_3x_3...+w_nx_n}{w_1+w_2+w_3+\cdots+w_n}$. Weighted values, such as $w_1, w_2, w_3, \dots w_n$ are assigned to each member of the set $x_1, x_2, x_3, \dots x_n$. When calculating the weighted mean, make sure a weight value for each member of the set is used.

Review Video: All About Averages
Visit mometrix.com/academy and enter code: 176521

MEDIAN

The **statistical median** is the value in the middle of the set of data. To find the median, list all data values in order from smallest to largest or from largest to smallest. Any value that is repeated in the set must be listed the number of times it appears. If there are an odd number of data values, the median is the value in the middle of the list. If there is an even number of data values, the median is the arithmetic mean of the two middle values.

For example, the median of the data set {88, 72, 61, 90, 97, 68, 88, 79, 86, 93, 97, 71, 80, 84, 88} is 86 since the ordered set is {61, 68, 71, 72, 79, 80, 84, **86**, 88, 88, 88, 90, 93, 97, 97}.

The big disadvantage of using the median as a measure of central tendency is that is relies solely on a value's relative size as compared to the other values in the set. When the individual values in a set of data are evenly dispersed, the median can be an accurate tool. However, if there is a group of rather large values or a group of rather small values that are not offset by a different group of values, the information that can be inferred from the median may not be accurate because the distribution of values is skewed.

MODE

The **statistical mode** is the data value that occurs the greatest number of times in the data set. It is possible to have exactly one mode, more than one mode, or no mode. To find the mode of a set of data, arrange the data like you do to find the median (all values in order, listing all multiples of data values). Count the number of times each value appears in the data set. If all values appear an equal number of times, there is no mode. If one value appears more than any other value, that value is the mode. If two or more values appear the same number of times, but there are other values that appear fewer times and no values that appear more times, all of those values are the modes.

For example, the mode of the data set {**88**, 72, 61, 90, 97, 68, **88**, 79, 86, 93, 97, 71, 80, 84, **88**} is 88.

The main disadvantage of the mode is that the values of the other data in the set have no bearing on the mode. The mode may be the largest value, the smallest value, or a value anywhere in between in the set. The mode only tells which value or values, if any, occurred the greatest number of times. It does not give any suggestions about the remaining values in the set.

Review Video: Mean, Median, and Mode
Visit mometrix.com/academy and enter code: 286207

Copyright © Mometrix Media. You have been licensed one copy of this document for personal use only. Any other reproduction or redistribution is strictly prohibited. All rights reserved.
This content is provided for test preparation purposes only and does not imply an endorsement by Mometrix of any particular political, scientific, or religious point of view.

Reading Comprehension

Transform passive reading into active learning! After immersing yourself in this chapter, put your comprehension to the test by taking a quiz. The insights you gained will stay with you longer this way. Scan the QR code to go directly to the chapter quiz interface for this study guide. If you're using a computer, simply visit the bonus page at **mometrix.com/bonus948/ibewapt** and click the Chapter Quizzes link.

Word Roots and Prefixes and Suffixes

Affixes

Affixes in the English language are morphemes that are added to words to create related but different words. Derivational affixes form new words based on and related to the original words. For example, the affix *-ness* added to the end of the adjective *happy* forms the noun *happiness.* Inflectional affixes form different grammatical versions of words. For example, the plural affix *-s* changes the singular noun *book* to the plural noun *books*, and the past tense affix *-ed* changes the present tense verb *look* to the past tense *looked.* Prefixes are affixes placed in front of words. For example, *heat* means to make hot; *preheat* means to heat in advance. Suffixes are affixes placed at the ends of words. The *happiness* example above contains the suffix *-ness.* Circumfixes add parts both before and after words, such as how *light* becomes *enlighten* with the prefix *en-* and the suffix *-en.* Interfixes create compound words via central affixes: *speed* and *meter* become *speedometer* via the interfix *-o-*.

Review Video: Affixes Visit mometrix.com/academy and enter code: 782422

Word Roots, Prefixes, and Suffixes to Help Determine Meanings of Words

Many English words were formed from combining multiple sources. For example, the Latin *habēre* means "to have," and the prefixes *in-* and *im-* mean a lack or prevention of something, as in *insufficient* and *imperfect.* Latin combined *in-* with *habēre* to form *inhibēre,* whose past participle was *inhibitus.* This is the origin of the English word *inhibit,* meaning to prevent from having. Hence by knowing the meanings of both the prefix and the root, one can decipher the word meaning. In Greek, the root *enkephalo-* refers to the brain. Many medical terms are based on this root, such as encephalitis and hydrocephalus. Understanding the prefix and suffix meanings (*-itis* means inflammation; *hydro-* means water) allows a person to deduce that encephalitis refers to brain inflammation and hydrocephalus refers to water (or other fluid) in the brain.

Review Video: Determining Word Meanings Visit mometrix.com/academy and enter code: 894894

Prefixes

Knowing common prefixes is helpful for all readers as they try to determining meanings or definitions of unfamiliar words. For example, a common word used when cooking is *preheat.* Knowing that *pre-* means in advance can also inform them that *presume* means to assume in advance, that *prejudice* means advance judgment, and that this understanding can be applied to many other words beginning with *pre-*. Knowing that the prefix *dis-* indicates opposition informs

Copyright © Mometrix Media. You have been licensed one copy of this document for personal use only. Any other reproduction or redistribution is strictly prohibited. All rights reserved.
This content is provided for test preparation purposes only and does not imply an endorsement by Mometrix of any particular political, scientific, or religious point of view.

the meanings of words like *disbar, disagree, disestablish,* and many more. Knowing *dys-* means bad, impaired, abnormal, or difficult informs *dyslogistic, dysfunctional, dysphagia,* and *dysplasia.*

SUFFIXES

In English, certain suffixes generally indicate both that a word is a noun, and that the noun represents a state of being or quality. For example, *-ness* is commonly used to change an adjective into its noun form, as with *happy* and *happiness, nice* and *niceness,* and so on. The suffix *–tion* is commonly used to transform a verb into its noun form, as with *converse* and *conversation or move* and *motion*. Thus, if readers are unfamiliar with the second form of a word, knowing the meaning of the transforming suffix can help them determine meaning.

PREFIXES FOR NUMBERS

Prefix	Definition	Examples
bi-	two	bisect, biennial
mono-	one, single	monogamy, monologue
poly-	many	polymorphous, polygamous
semi-	half, partly	semicircle, semicolon
uni-	one	uniform, unity

Copyright © Mometrix Media. You have been licensed one copy of this document for personal use only. Any other reproduction or redistribution is strictly prohibited. All rights reserved.
This content is provided for test preparation purposes only and does not imply an endorsement by Mometrix of any particular political, scientific, or religious point of view.

PREFIXES FOR TIME, DIRECTION, AND SPACE

Prefix	Definition	Examples
a-	in, on, of, up, to	abed, afoot
ab-	from, away, off	abdicate, abjure
ad-	to, toward	advance, adventure
ante-	before, previous	antecedent, antedate
anti-	against, opposing	antipathy, antidote
cata-	down, away, thoroughly	catastrophe, cataclysm
circum-	around	circumspect, circumference
com-	with, together, very	commotion, complicate
contra-	against, opposing	contradict, contravene
de-	from	depart
dia-	through, across, apart	diameter, diagnose
dis-	away, off, down, not	dissent, disappear
epi-	upon	epilogue
ex-	out	extract, excerpt
hypo-	under, beneath	hypodermic, hypothesis
inter-	among, between	intercede, interrupt
intra-	within	intramural, intrastate
ob-	against, opposing	objection
per-	through	perceive, permit
peri-	around	periscope, perimeter
post-	after, following	postpone, postscript
pre-	before, previous	prevent, preclude
pro-	forward, in place of	propel, pronoun
retro-	back, backward	retrospect, retrograde
sub-	under, beneath	subjugate, substitute
super-	above, extra	supersede, supernumerary
trans-	across, beyond, over	transact, transport
ultra-	beyond, excessively	ultramodern, ultrasonic

NEGATIVE PREFIXES

Prefix	Definition	Examples
a-	without, lacking	atheist, agnostic
in-	not, opposing	incapable, ineligible
non-	not	nonentity, nonsense
un-	not, reverse of	unhappy, unlock

Copyright © Mometrix Media. You have been licensed one copy of this document for personal use only. Any other reproduction or redistribution is strictly prohibited. All rights reserved.
This content is provided for test preparation purposes only and does not imply an endorsement by Mometrix of any particular political, scientific, or religious point of view.

Extra Prefixes

Prefix	Definition	Examples
for-	away, off, from	forget, forswear
fore-	previous	foretell, forefathers
homo-	same, equal	homogenized, homonym
hyper-	excessive, over	hypercritical, hypertension
in-	in, into	intrude, invade
mal-	bad, poorly, not	malfunction, malpractice
mis-	bad, poorly, not	misspell, misfire
neo-	new	Neolithic, neoconservative
omni-	all, everywhere	omniscient, omnivore
ortho-	right, straight	orthogonal, orthodox
over-	above	overbearing, oversight
pan-	all, entire	panorama, pandemonium
para-	beside, beyond	parallel, paradox
re-	backward, again	revoke, recur
sym-	with, together	sympathy, symphony

Below is a list of common suffixes and their meanings:

Adjective Suffixes

Suffix	Definition	Examples
-able (-ible)	capable of being	toler*able*, ed*ible*
-esque	in the style of, like	picturesque, grotesque
-ful	filled with, marked by	thankful, zestful
-ific	make, cause	terrific, beatific
-ish	suggesting, like	churlish, childish
-less	lacking, without	hopeless, countless
-ous	marked by, given to	religious, riotous

Copyright © Mometrix Media. You have been licensed one copy of this document for personal use only. Any other reproduction or redistribution is strictly prohibited. All rights reserved.
This content is provided for test preparation purposes only and does not imply an endorsement by Mometrix of any particular political, scientific, or religious point of view.

Noun Suffixes

Suffix	Definition	Examples
-acy	state, condition	accuracy, privacy
-ance	act, condition, fact	acceptance, vigilance
-ard	one that does excessively	drunkard, sluggard
-ation	action, state, result	occupation, starvation
-dom	state, rank, condition	serfdom, wisdom
-er (-or)	office, action	teach*er*, elevat*or*, hon*or*
-ess	feminine	waitress, duchess
-hood	state, condition	manhood, statehood
-ion	action, result, state	union, fusion
-ism	act, manner, doctrine	barbarism, socialism
-ist	worker, follower	monopolist, socialist
-ity (-ty)	state, quality, condition	acid*ity*, civil*ity*, twen*ty*
-ment	result, action	Refreshment
-ness	quality, state	greatness, tallness
-ship	position	internship, statesmanship
-sion (-tion)	state, result	revi*sion*, expedi*tion*
-th	act, state, quality	warmth, width
-tude	quality, state, result	magnitude, fortitude

Verb Suffixes

Suffix	Definition	Examples
-ate	having, showing	separate, desolate
-en	cause to be, become	deepen, strengthen
-fy	make, cause to have	glorify, fortify
-ize	cause to be, treat with	sterilize, mechanize

Nuance and Word Meanings

Synonyms and Antonyms

When you understand how words relate to each other, you will discover more in a passage. This is explained by understanding **synonyms** (e.g., words that mean the same thing) and **antonyms** (e.g., words that mean the opposite of one another). As an example, *dry* and *arid* are synonyms, and *dry* and *wet* are antonyms.

There are many pairs of words in English that can be considered synonyms, despite having slightly different definitions. For instance, the words *friendly* and *collegial* can both be used to describe a warm interpersonal relationship, and one would be correct to call them synonyms. However, *collegial* (kin to *colleague*) is often used in reference to professional or academic relationships, and *friendly* has no such connotation.

If the difference between the two words is too great, then they should not be called synonyms. *Hot* and *warm* are not synonyms because their meanings are too distinct. A good way to determine whether two words are synonyms is to substitute one word for the other word and verify that the meaning of the sentence has not changed. Substituting *warm* for *hot* in a sentence would convey a different meaning. Although warm and hot may seem close in meaning, warm generally means that the temperature is moderate, and hot generally means that the temperature is excessively high.

Copyright © Mometrix Media. You have been licensed one copy of this document for personal use only. Any other reproduction or redistribution is strictly prohibited. All rights reserved.
This content is provided for test preparation purposes only and does not imply an endorsement by Mometrix of any particular political, scientific, or religious point of view.

Antonyms are words with opposite meanings. *Light* and *dark*, *up* and *down*, *right* and *left*, *good* and *bad*: these are all sets of antonyms. Be careful to distinguish between antonyms and pairs of words that are simply different. *Black* and *gray*, for instance, are not antonyms because gray is not the opposite of black. *Black* and *white*, on the other hand, are antonyms.

Not every word has an antonym. For instance, many nouns do not. What would be the antonym of *chair*? During your exam, the questions related to antonyms are more likely to concern adjectives. You will recall that adjectives are words that describe a noun. Some common adjectives include *purple*, *fast*, *skinny*, and *sweet*. From those four adjectives, *purple* is the item that lacks a group of obvious antonyms.

> **Review Video: Synonyms and Antonyms**
> Visit mometrix.com/academy and enter code: 105612

Denotative vs. Connotative Meaning

The **denotative** meaning of a word is the literal meaning. The **connotative** meaning goes beyond the denotative meaning to include the emotional reaction that a word may invoke. The connotative meaning often takes the denotative meaning a step further due to associations the reader makes with the denotative meaning. Readers can differentiate between the denotative and connotative meanings by first recognizing how authors use each meaning. Most non-fiction, for example, is fact-based and authors do not use flowery, figurative language. The reader can assume that the writer is using the denotative meaning of words. In fiction, the author may use the connotative meaning. Readers can determine whether the author is using the denotative or connotative meaning of a word by implementing context clues.

> **Review Video: Connotation and Denotation**
> Visit mometrix.com/academy and enter code: 310092

Nuances of Word Meaning Relative to Connotation, Denotation, Diction, and Usage

A word's denotation is simply its objective dictionary definition. However, its connotation refers to the subjective associations, often emotional, that specific words evoke in listeners and readers. Two or more words can have the same dictionary meaning, but very different connotations. Writers use diction (word choice) to convey various nuances of thought and emotion by selecting synonyms for other words that best communicate the associations they want to trigger for readers. For example, a car engine is naturally greasy; in this sense, "greasy" is a neutral term. But when a person's smile, appearance, or clothing is described as "greasy," it has a negative connotation. Some words have even gained additional or different meanings over time. For example, *awful* used to be used to describe things that evoked a sense of awe. When *awful* is separated into its root word, awe, and suffix, -ful, it can be understood to mean "full of awe." However, the word is now commonly used to describe things that evoke repulsion, terror, or another intense, negative reaction.

> **Review Video: Word Usage in Sentences**
> Visit mometrix.com/academy and enter code: 197863

Copyright © Mometrix Media. You have been licensed one copy of this document for personal use only. Any other reproduction or redistribution is strictly prohibited. All rights reserved.
This content is provided for test preparation purposes only and does not imply an endorsement by Mometrix of any particular political, scientific, or religious point of view.

Using Context to Determine Meaning

Context Clues

Readers of all levels will encounter words that they have either never seen or have encountered only on a limited basis. The best way to define a word in **context** is to look for nearby words that can assist in revealing the meaning of the word. For instance, unfamiliar nouns are often accompanied by examples that provide a definition. Consider the following sentence: *Dave arrived at the party in hilarious garb: a leopard-print shirt, buckskin trousers, and bright green sneakers.* If a reader was unfamiliar with the meaning of garb, he or she could read the examples (i.e., a leopard-print shirt, buckskin trousers, and bright green sneakers) and quickly determine that the word means *clothing*. Examples will not always be this obvious. Consider this sentence: *Parsley, lemon, and flowers were just a few of the items he used as garnishes.* Here, the word *garnishes* is exemplified by parsley, lemon, and flowers. Readers who have eaten in a variety of restaurants will probably be able to identify a garnish as something used to decorate a plate.

> **Review Video: Reading Comprehension: Using Context Clues**
> Visit mometrix.com/academy and enter code: 613660

Using Contrast in Context Clues

In addition to looking at the context of a passage, readers can use contrast to define an unfamiliar word in context. In many sentences, the author will not describe the unfamiliar word directly; instead, he or she will describe the opposite of the unfamiliar word. Thus, you are provided with some information that will bring you closer to defining the word. Consider the following example: *Despite his intelligence, Hector's low brow and bad posture made him look obtuse.* The author writes that Hector's appearance does not convey intelligence. Therefore, *obtuse* must mean unintelligent. Here is another example: *Despite the horrible weather, we were beatific about our trip to Alaska.* The word *despite* indicates that the speaker's feelings were at odds with the weather. Since the weather is described as *horrible*, then *beatific* must mean something positive.

Substitution to Find Meaning

In some cases, there will be very few contextual clues to help a reader define the meaning of an unfamiliar word. When this happens, one strategy that readers may employ is **substitution**. A good reader will brainstorm some possible synonyms for the given word, and he or she will substitute these words into the sentence. If the sentence and the surrounding passage continue to make sense, then the substitution has revealed at least some information about the unfamiliar word. Consider the sentence: *Frank's admonition rang in her ears as she climbed the mountain.* A reader unfamiliar with *admonition* might come up with some substitutions like *vow, promise, advice, complaint*, or *compliment*. All of these words make general sense of the sentence, though their meanings are diverse. However, this process has suggested that an admonition is some sort of message. The substitution strategy is rarely able to pinpoint a precise definition, but this process can be effective as a last resort.

Occasionally, you will be able to define an unfamiliar word by looking at the descriptive words in the context. Consider the following sentence: *Fred dragged the recalcitrant boy kicking and screaming up the stairs.* The words *dragged, kicking*, and *screaming* all suggest that the boy does not want to go up the stairs. The reader may assume that *recalcitrant* means something like unwilling or protesting. In this example, an unfamiliar adjective was identified.

Additionally, using description to define an unfamiliar noun is a common practice compared to unfamiliar adjectives, as in this sentence: *Don's wrinkled frown and constantly shaking fist identified*

Copyright © Mometrix Media. You have been licensed one copy of this document for personal use only. Any other reproduction or redistribution is strictly prohibited. All rights reserved.
This content is provided for test preparation purposes only and does not imply an endorsement by Mometrix of any particular political, scientific, or religious point of view.

him as a curmudgeon of the first order. Don is described as having a *wrinkled frown and constantly shaking fist*, suggesting that a *curmudgeon* must be a grumpy person. Contrasts do not always provide detailed information about the unfamiliar word, but they at least give the reader some clues.

Words with Multiple Meanings

When a word has more than one meaning, readers can have difficulty determining how the word is being used in a given sentence. For instance, the verb *cleave*, can mean either *join* or *separate*. When readers come upon this word, they will have to select the definition that makes the most sense. Consider the following sentence: *Hermione's knife cleaved the bread cleanly*. Since a knife cannot join bread together, the word must indicate separation. A slightly more difficult example would be the sentence: *The birds cleaved to one another as they flew from the oak tree.* Immediately, the presence of the words *to one another* should suggest that in this sentence *cleave* is being used to mean *join*. Discovering the intent of a word with multiple meanings requires the same tricks as defining an unknown word: look for contextual clues and evaluate the substituted words.

Context Clues to Help Determine Meanings of Words

If readers simply bypass unknown words, they can reach unclear conclusions about what they read. However, looking for the definition of every unfamiliar word in the dictionary can slow their reading progress. Moreover, the dictionary may list multiple definitions for a word, so readers must search the word's context for meaning. Hence context is important to new vocabulary regardless of reader methods. Four types of context clues are examples, definitions, descriptive words, and opposites. Authors may use a certain word, and then follow it with several different examples of what it describes. Sometimes authors actually supply a definition of a word they use, which is especially true in informational and technical texts. Authors may use descriptive words that elaborate upon a vocabulary word they just used. Authors may also use opposites with negation that help define meaning.

Examples and Definitions

An author may use a word and then give examples that illustrate its meaning. Consider this text: "Teachers who do not know how to use sign language can help students who are deaf or hard of hearing understand certain instructions by using gestures instead, like pointing their fingers to indicate which direction to look or go; holding up a hand, palm outward, to indicate stopping; holding the hands flat, palms up, curling a finger toward oneself in a beckoning motion to indicate 'come here'; or curling all fingers toward oneself repeatedly to indicate 'come on', 'more', or 'continue.'" The author of this text has used the word "gestures" and then followed it with examples, so a reader unfamiliar with the word could deduce from the examples that "gestures" means "hand motions." Readers can find examples by looking for signal words "for example," "for instance," "like," "such as," and "e.g."

While readers sometimes have to look for definitions of unfamiliar words in a dictionary or do some work to determine a word's meaning from its surrounding context, at other times an author may make it easier for readers by defining certain words. For example, an author may write, "The company did not have sufficient capital, that is, available money, to continue operations." The author defined "capital" as "available money," and heralded the definition with the phrase "that is." Another way that authors supply word definitions is with appositives. Rather than being introduced by a signal phrase like "that is," "namely," or "meaning," an appositive comes after the vocabulary word it defines and is enclosed within two commas. For example, an author may write, "The Indians introduced the Pilgrims to pemmican, cakes they made of lean meat dried and mixed with fat, which

Copyright © Mometrix Media. You have been licensed one copy of this document for personal use only. Any other reproduction or redistribution is strictly prohibited. All rights reserved.
This content is provided for test preparation purposes only and does not imply an endorsement by Mometrix of any particular political, scientific, or religious point of view.

proved greatly beneficial to keep settlers from starving while trapping." In this example, the appositive phrase following "pemmican" and preceding "which" defines the word "pemmican."

Descriptions

When readers encounter a word they do not recognize in a text, the author may expand on that word to illustrate it better. While the author may do this to make the prose more picturesque and vivid, the reader can also take advantage of this description to provide context clues to the meaning of the unfamiliar word. For example, an author may write, "The man sitting next to me on the airplane was obese. His shirt stretched across his vast expanse of flesh, strained almost to bursting." The descriptive second sentence elaborates on and helps to define the previous sentence's word "obese" to mean extremely fat. A reader unfamiliar with the word "repugnant" can decipher its meaning through an author's accompanying description: "The way the child grimaced and shuddered as he swallowed the medicine showed that its taste was particularly repugnant."

Opposites

Text authors sometimes introduce a contrasting or opposing idea before or after a concept they present. They may do this to emphasize or heighten the idea they present by contrasting it with something that is the reverse. However, readers can also use these context clues to understand familiar words. For example, an author may write, "Our conversation was not cheery. We sat and talked very solemnly about his experience and a number of similar events." The reader who is not familiar with the word "solemnly" can deduce by the author's preceding use of "not cheery" that "solemn" means the opposite of cheery or happy, so it must mean serious or sad. Or if someone writes, "Don't condemn his entire project because you couldn't find anything good to say about it," readers unfamiliar with "condemn" can understand from the sentence structure that it means the opposite of saying anything good, so it must mean reject, dismiss, or disapprove. "Entire" adds another context clue, meaning total or complete rejection.

Syntax to Determine Part of Speech and Meanings of Words

Syntax refers to sentence structure and word order. Suppose that a reader encounters an unfamiliar word when reading a text. To illustrate, consider an invented word like "splunch." If this word is used in a sentence like "Please splunch that ball to me," the reader can assume from syntactic context that "splunch" is a verb. We would not use a noun, adjective, adverb, or preposition with the object "that ball," and the prepositional phrase "to me" further indicates "splunch" represents an action. However, in the sentence, "Please hand that splunch to me," the reader can assume that "splunch" is a noun. Demonstrative adjectives like "that" modify nouns. Also, we hand someone some*thing*—a thing being a noun; we do not hand someone a verb, adjective, or adverb. Some sentences contain further clues. For example, from the sentence, "The princess wore the glittering splunch on her head," the reader can deduce that it is a crown, tiara, or something similar from the syntactic context, without knowing the word.

Syntax to Indicate Different Meanings of Similar Sentences

The syntax, or structure, of a sentence affords grammatical cues that aid readers in comprehending the meanings of words, phrases, and sentences in the texts that they read. Seemingly minor differences in how the words or phrases in a sentence are ordered can make major differences in meaning. For example, two sentences can use exactly the same words but have different meanings based on the word order:

- "The man with a broken arm sat in a chair."
- "The man sat in a chair with a broken arm."

Copyright © Mometrix Media. You have been licensed one copy of this document for personal use only. Any other reproduction or redistribution is strictly prohibited. All rights reserved.
This content is provided for test preparation purposes only and does not imply an endorsement by Mometrix of any particular political, scientific, or religious point of view.

While both sentences indicate that a man sat in a chair, differing syntax indicates whether the man's or chair's arm was broken.

Review Video: Syntax
Visit mometrix.com/academy and enter code: 242280

Determining Meaning of Phrases and Paragraphs

Like unknown words, the meanings of phrases, paragraphs, and entire works can also be difficult to discern. Each of these can be better understood with added context. However, for larger groups of words, more context is needed. Unclear phrases are similar to unclear words, and the same methods can be used to understand their meaning. However, it is also important to consider how the individual words in the phrase work together. Paragraphs are a bit more complicated. Just as words must be compared to other words in a sentence, paragraphs must be compared to other paragraphs in a composition or a section.

Determining Meaning in Various Types of Compositions

To understand the meaning of an entire composition, the type of composition must be considered. **Expository writing** is generally organized so that each paragraph focuses on explaining one idea, or part of an idea, and its relevance. **Persuasive writing** uses paragraphs for different purposes to organize the parts of the argument. **Unclear paragraphs** must be read in the context of the paragraphs around them for their meaning to be fully understood. The meaning of full texts can also be unclear at times. The purpose of composition is also important for understanding the meaning of a text. To quickly understand the broad meaning of a text, look to the introductory and concluding paragraphs. Fictional texts are different. Some fictional works have implicit meanings, but some do not. The target audience must be considered for understanding texts that do have an implicit meaning, as most children's fiction will clearly state any lessons or morals. For other fiction, the application of literary theories and criticism may be helpful for understanding the text.

Main Ideas and Supporting Details

Identifying Topics and Main Ideas

One of the most important skills in reading comprehension is the identification of **topics** and **main ideas**. There is a subtle difference between these two features. The topic is the subject of a text (i.e., what the text is all about). The main idea, on the other hand, is the most important point being made by the author. The topic is usually expressed in a few words at the most while the main idea often needs a full sentence to be completely defined. As an example, a short passage might be written on the topic of penguins, and the main idea could be written as *Penguins are different from other birds in many ways*. In most nonfiction writing, the topic and the main idea will be **stated directly** and often appear in a sentence at the very beginning or end of the text. When being tested on an understanding of the author's topic, you may be able to skim the passage for the general idea by reading only the first sentence of each paragraph. A body paragraph's first sentence is often—but not always—the main **topic sentence** which gives you a summary of the content in the paragraph.

However, there are cases in which the reader must figure out an **unstated** topic or main idea. In these instances, you must read every sentence of the text and try to come up with an overarching idea that is supported by each of those sentences.

Copyright © Mometrix Media. You have been licensed one copy of this document for personal use only. Any other reproduction or redistribution is strictly prohibited. All rights reserved.
This content is provided for test preparation purposes only and does not imply an endorsement by Mometrix of any particular political, scientific, or religious point of view.

Note: The main idea should not be confused with the thesis statement. While the main idea gives a brief, general summary of a text, the thesis statement provides a **specific perspective** on an issue that the author supports with evidence.

Review Video: Topics and Main Ideas
Visit mometrix.com/academy and enter code: 407801

SUPPORTING DETAILS

Supporting details are smaller pieces of evidence that provide backing for the main point. In order to show that a main idea is correct or valid, an author must add details that prove their point. All texts contain details, but they are only classified as supporting details when they serve to reinforce some larger point. Supporting details are most commonly found in informative and persuasive texts. In some cases, they will be clearly indicated with terms like *for example* or *for instance*, or they will be enumerated with terms like *first*, *second*, and *last*. However, you need to be prepared for texts that do not contain those indicators. As a reader, you should consider whether the author's supporting details really back up his or her main point. Details can be factual and correct, yet they may not be **relevant** to the author's point. Conversely, details can be relevant, but be ineffective because they are based on opinion or assertions that cannot be proven.

Review Video: Supporting Details
Visit mometrix.com/academy and enter code: 396297

Author's Purpose

AUTHOR'S PURPOSE

Usually, identifying the author's **purpose** is easier than identifying his or her **position**. In most cases, the author has no interest in hiding his or her purpose. A text that is meant to entertain, for instance, should be written to please the reader. Most narratives, or stories, are written to entertain, though they may also inform or persuade. Informative texts are easy to identify, while the most difficult purpose of a text to identify is persuasion because the author has an interest in making this purpose hard to detect. When a reader discovers that the author is trying to persuade, he or she should be skeptical of the argument. For this reason, persuasive texts often try to establish an entertaining tone and hope to amuse the reader into agreement. On the other hand, an informative tone may be implemented to create an appearance of authority and objectivity.

An author's purpose is evident often in the **organization** of the text (e.g., section headings in bold font points to an informative text). However, you may not have such organization available to you in your exam. Instead, if the author makes his or her main idea clear from the beginning, then the likely purpose of the text is to **inform**. If the author begins by making a claim and provides various arguments to support that claim, then the purpose is probably to **persuade**. If the author tells a story or wants to gain the reader's attention more than to push a particular point or deliver information, then his or her purpose is most likely to **entertain**. As a reader, you must judge authors on how well they accomplish their purpose. In other words, you need to consider the type of passage (e.g., technical, persuasive, etc.) that the author has written and if the author has followed the requirements of the passage type.

Review Video: Understanding the Author's Intent
Visit mometrix.com/academy and enter code: 511819

Copyright © Mometrix Media. You have been licensed one copy of this document for personal use only. Any other reproduction or redistribution is strictly prohibited. All rights reserved.
This content is provided for test preparation purposes only and does not imply an endorsement by Mometrix of any particular political, scientific, or religious point of view.

Informational Texts

An **informational text** is written to educate and enlighten readers. Informational texts are almost always nonfiction and are rarely structured as a story. The intention of an informational text is to deliver information in the most comprehensible way. So, look for the structure of the text to be very clear. In an informational text, the thesis statement is one or two sentences that normally appears at the end of the first paragraph. The author may use some colorful language, but he or she is likely to put more emphasis on clarity and precision. Informational essays do not typically appeal to the emotions. They often contain facts and figures and rarely include the opinion of the author; however, readers should remain aware of the possibility for bias as those facts are presented. Sometimes a persuasive essay can resemble an informative essay, especially if the author maintains an even tone and presents his or her views as if they were established fact.

Review Video: Informational Text
Visit mometrix.com/academy and enter code: 924964

Persuasive Writing

In a persuasive essay, the author is attempting to change the reader's mind or **convince** him or her of something that he or she did not believe previously. There are several identifying characteristics of **persuasive writing**. One is **opinion presented as fact**. When authors attempt to persuade readers, they often present their opinions as if they were fact. Readers must be on guard for statements that sound factual but which cannot be subjected to research, observation, or experiment. Another characteristic of persuasive writing is **emotional language**. An author will often try to play on the emotions of readers by appealing to their sympathy or sense of morality. When an author uses colorful or evocative language with the intent of arousing the reader's passions, then the author may be attempting to persuade. Finally, in many cases, a persuasive text will give an **unfair explanation of opposing positions**, if these positions are mentioned at all.

Entertaining Texts

The success or failure of an author's intent to **entertain** is determined by those who read the author's work. Entertaining texts may be either fiction or nonfiction, and they may describe real or imagined people, places, and events. Entertaining texts are often narratives or poems. A text that is written to entertain is likely to contain **colorful language** that engages the imagination and the emotions. Such writing often features a great deal of figurative language, which typically enlivens the subject matter with images and analogies.

Though an entertaining text is not usually written to persuade or inform, authors may accomplish both of these tasks in their work. An entertaining text may *appeal to the reader's emotions* and cause him or her to think differently about a particular subject. In any case, entertaining texts tend to showcase the personality of the author more than other types of writing.

Descriptive Text

In a sense, almost all writing is descriptive, insofar as an author seeks to describe events, ideas, or people to the reader. Some texts, however, are primarily concerned with **description**. A descriptive text focuses on a particular subject and attempts to depict the subject in a way that will be clear to readers. Descriptive texts contain many adjectives and adverbs (i.e., words that give shades of meaning and create a more detailed mental picture for the reader). A descriptive text fails when it is unclear to the reader. A descriptive text will certainly be informative and may be persuasive and entertaining as well.

Copyright © Mometrix Media. You have been licensed one copy of this document for personal use only. Any other reproduction or redistribution is strictly prohibited. All rights reserved.
This content is provided for test preparation purposes only and does not imply an endorsement by Mometrix of any particular political, scientific, or religious point of view.

Review Video: Descriptive Texts
Visit mometrix.com/academy and enter code: 174903

Expression of Feelings

When an author intends to **express feelings**, he or she may use **expressive and bold language**. An author may write with emotion for any number of reasons. Sometimes, authors will express feelings because they are describing a personal situation of great pain or happiness. In other situations, authors will attempt to persuade the reader and will use emotion to stir up the passions. This kind of expression is easy to identify when the writer uses phrases like *I felt* and *I sense*. However, readers may find that the author will simply describe feelings without introducing them. As a reader, you must know the importance of recognizing when an author is expressing emotion and not to become overwhelmed by sympathy or passion. Readers should maintain some **detachment** so that they can still evaluate the strength of the author's argument or the quality of the writing.

Review Video: Emotional Language in Literature
Visit mometrix.com/academy and enter code: 759390

Expository Passage

An **expository** passage aims to **inform** and enlighten readers. Expository passages are nonfiction and usually center around a simple, easily defined topic. Since the goal of exposition is to teach, such a passage should be as clear as possible. Often, an expository passage contains helpful organizing words, like *first*, *next*, *for example*, and *therefore*. These words keep the reader **oriented** in the text. Although expository passages do not need to feature colorful language and artful writing, they are often more effective with these features. For a reader, the challenge of expository passages is to maintain steady attention. Expository passages are not always about subjects that will naturally interest a reader, so the writer is often more concerned with **clarity** and **comprehensibility** than with engaging the reader. By reading actively, you can ensure a good habit of focus when reading an expository passage.

Review Video: Expository Passages
Visit mometrix.com/academy and enter code: 256515

Narrative Passage

A **narrative** passage is a story that can be fiction or nonfiction. However, there are a few elements that a text must have in order to be classified as a narrative. First, the text must have a **plot** (i.e., a series of events). Narratives often proceed in a clear sequence, but this is not a requirement. If the narrative is good, then these events will be interesting to readers. Second, a narrative has **characters**. These characters could be people, animals, or even inanimate objects—so long as they participate in the plot. Third, a narrative passage often contains **figurative language** which is meant to stimulate the imagination of readers by making comparisons and observations. For instance, a *metaphor*, a common piece of figurative language, is a description of one thing in terms of another. *The moon was a frosty snowball* is an example of a metaphor. In the literal sense this is obviously untrue, but the comparison suggests a certain mood for the reader.

Technical Passage

A **technical** passage is written to *describe* a complex object or process. Technical writing is common in medical and technological fields, in which complex ideas of mathematics, science, and engineering need to be explained *simply* and *clearly*. To ease comprehension, a technical passage

Copyright © Mometrix Media. You have been licensed one copy of this document for personal use only. Any other reproduction or redistribution is strictly prohibited. All rights reserved.
This content is provided for test preparation purposes only and does not imply an endorsement by Mometrix of any particular political, scientific, or religious point of view.

usually proceeds in a very logical order. Technical passages often have clear headings and subheadings, which are used to keep the reader oriented in the text. Additionally, you will find that these passages divide sections up with numbers or letters. Many technical passages look more like an outline than a piece of prose. The amount of **jargon** or difficult vocabulary will vary in a technical passage depending on the intended audience. As much as possible, technical passages try to avoid language that the reader will have to research in order to understand the message, yet readers will find that jargon cannot always be avoided.

Review Video: Technical Passages
Visit mometrix.com/academy and enter code: 478923

Common Organizations of Texts

Organization of the Text

The way a text is organized can help readers understand the author's intent and his or her conclusions. There are various ways to organize a text, and each one has a purpose and use. Usually, authors will organize information logically in a passage so the reader can follow and locate the information within the text. However, since not all passages are written with the same logical structure, you need to be familiar with several different types of passage structure.

Review Video: Organizational Methods to Structure Text
Visit mometrix.com/academy and enter code: 606263

Review Video: Sequence of Events in a Story
Visit mometrix.com/academy and enter code: 807512

Chronological

When using **chronological** order, the author presents information in the order that it happened. For example, biographies are typically written in chronological order. The subject's birth and childhood are presented first, followed by their adult life, and lastly the events leading up to the person's death.

Cause and Effect

One of the most common text structures is **cause and effect**. A **cause** is an act or event that makes something happen, and an **effect** is the thing that happens as a result of the cause. A cause-and-effect relationship is not always explicit, but there are some terms in English that signal causes, such as *since*, *because*, and *due to*. Furthermore, terms that signal effects include *consequently, therefore, this leads to*. As an example, consider the sentence *Because the sky was clear, Ron did not bring an umbrella*. The cause is the clear sky, and the effect is that Ron did not bring an umbrella. However, readers may find that sometimes the cause-and-effect relationship will not be clearly noted. For instance, the sentence *He was late and missed the meeting* does not contain any signaling words, but the sentence still contains a cause (he was late) and an effect (he missed the meeting).

Review Video: Cause and Effect
Visit mometrix.com/academy and enter code: 868099

Review Video: Rhetorical Strategy of Cause and Effect Analysis
Visit mometrix.com/academy and enter code: 725944

Copyright © Mometrix Media. You have been licensed one copy of this document for personal use only. Any other reproduction or redistribution is strictly prohibited. All rights reserved.
This content is provided for test preparation purposes only and does not imply an endorsement by Mometrix of any particular political, scientific, or religious point of view.

Multiple Effects

Be aware of the possibility for a single cause to have **multiple effects.** (e.g., *Single cause*: Because you left your homework on the table, your dog engulfed the assignment. *Multiple effects*: As a result, you receive a failing grade, your parents do not allow you to go out with your friends, you miss out on the new movie, and one of your classmates spoils it for you before you have another chance to watch it).

Multiple Causes

Also, there is the possibility for a single effect to have **multiple causes.** (e.g., *Single effect*: Alan has a fever. *Multiple causes*: An unexpected cold front came through the area, and Alan forgot to take his multi-vitamin to avoid getting sick.) Additionally, an effect can in turn be the cause of another effect, in what is known as a cause-and-effect chain. (e.g., As a result of her disdain for procrastination, Lynn prepared for her exam. This led to her passing her test with high marks. Hence, her resume was accepted and her application was approved.)

Cause and Effect in Persuasive Essays

Persuasive essays, in which an author tries to make a convincing argument and change the minds of readers, usually include cause-and-effect relationships. However, these relationships should not always be taken at face value. Frequently, an author will assume a cause or take an effect for granted. To read a persuasive essay effectively, readers need to judge the cause-and-effect relationships that the author is presenting. For instance, imagine an author wrote the following: *The parking deck has been unprofitable because people would prefer to ride their bikes.* The relationship is clear: the cause is that people prefer to ride their bikes, and the effect is that the parking deck has been unprofitable. However, readers should consider whether this argument is conclusive. Perhaps there are other reasons for the failure of the parking deck: a down economy, excessive fees, etc. Too often, authors present causal relationships as if they are fact rather than opinion. Readers should be on the alert for these dubious claims.

Problem-Solution

Some nonfiction texts are organized to **present a problem** followed by a solution. For this type of text, the problem is often explained before the solution is offered. In some cases, as when the problem is well known, the solution may be introduced briefly at the beginning. Other passages may focus on the solution, and the problem will be referenced only occasionally. Some texts will outline multiple solutions to a problem, leaving readers to choose among them. If the author has an interest or an allegiance to one solution, he or she may fail to mention or describe accurately some of the other solutions. Readers should be careful of the author's agenda when reading a problem-solution text. Only by understanding the author's perspective and interests can one develop a proper judgment of the proposed solution.

Compare and Contrast

Many texts follow the **compare-and-contrast** model in which the similarities and differences between two ideas or things are explored. Analysis of the similarities between ideas is called **comparison**. In an ideal comparison, the author places ideas or things in an equivalent structure, i.e., the author presents the ideas in the same way. If an author wants to show the similarities between cricket and baseball, then he or she may do so by summarizing the equipment and rules for each game. Be mindful of the similarities as they appear in the passage and take note of any

Copyright © Mometrix Media. You have been licensed one copy of this document for personal use only. Any other reproduction or redistribution is strictly prohibited. All rights reserved.
This content is provided for test preparation purposes only and does not imply an endorsement by Mometrix of any particular political, scientific, or religious point of view.

differences that are mentioned. Often, these small differences will only reinforce the more general similarity.

Review Video: Compare and Contrast
Visit mometrix.com/academy and enter code: 798319

Thinking critically about ideas and conclusions can seem like a daunting task. One way to ease this task is to understand the basic elements of ideas and writing techniques. Looking at the ways different ideas relate to each other can be a good way for readers to begin their analysis. For instance, sometimes authors will write about two ideas that are in opposition to each other. Or, one author will provide his or her ideas on a topic, and another author may respond in opposition. The analysis of these opposing ideas is known as **contrast**. Contrast is often marred by the author's obvious partiality to one of the ideas. A discerning reader will be put off by an author who does not engage in a fair fight. In an analysis of opposing ideas, both ideas should be presented in clear and reasonable terms. If the author does prefer a side, you need to read carefully to determine the areas where the author shows or avoids this preference. In an analysis of opposing ideas, you should proceed through the passage by marking the major differences point by point with an eye that is looking for an explanation of each side's view. For instance, in an analysis of capitalism and communism, there is an importance in outlining each side's view on labor, markets, prices, personal responsibility, etc. Additionally, as you read through the passages, you should note whether the opposing views present each side in a similar manner.

Sequence

Readers must be able to identify a text's **sequence**, or the order in which things happen. Often, when the sequence is very important to the author, the text is indicated with signal words like *first*, *then*, *next*, and *last*. However, a sequence can be merely implied and must be noted by the reader. Consider the sentence *He walked through the garden and gave water and fertilizer to the plants.* Clearly, the man did not walk through the garden before he collected water and fertilizer for the plants. So, the implied sequence is that he first collected water, then he collected fertilizer, next he walked through the garden, and last he gave water or fertilizer as necessary to the plants. Texts do not always proceed in an orderly sequence from first to last. Sometimes they begin at the end and start over at the beginning. As a reader, you can enhance your understanding of the passage by taking brief notes to clarify the sequence.

Review Video: Sequence
Visit mometrix.com/academy and enter code: 489027

Making and Evaluating Predictions

Making Predictions

When we read literature, **making predictions** about what will happen in the writing reinforces our purpose for reading and prepares us mentally. A **prediction** is a guess about what will happen next. Readers constantly make predictions based on what they have read and what they already know. We can make predictions before we begin reading and during our reading. Consider the following sentence: *Staring at the computer screen in shock, Kim blindly reached over for the brimming glass of water on the shelf to her side.* The sentence suggests that Kim is distracted, and that she is not looking at the glass that she is going to pick up. So, a reader might predict that Kim is going to knock over the glass. Of course, not every prediction will be accurate: perhaps Kim will pick the glass up

Copyright © Mometrix Media. You have been licensed one copy of this document for personal use only. Any other reproduction or redistribution is strictly prohibited. All rights reserved.
This content is provided for test preparation purposes only and does not imply an endorsement by Mometrix of any particular political, scientific, or religious point of view.

cleanly. Nevertheless, the author has certainly created the expectation that the water might be spilled.

As we read on, we can test the accuracy of our predictions, revise them in light of additional reading, and confirm or refute our predictions. Predictions are always subject to revision as the reader acquires more information. A reader can make predictions by observing the title and illustrations; noting the structure, characters, and subject; drawing on existing knowledge relative to the subject; and asking "why" and "who" questions. Connecting reading to what we already know enables us to learn new information and construct meaning. For example, before third-graders read a book about Johnny Appleseed, they may start a KWL chart—a list of what they *Know*, what they *Want* to know or learn, and what they have *Learned* after reading. Activating existing background knowledge and thinking about the text before reading improves comprehension.

Review Video: Predictive Reading
Visit mometrix.com/academy and enter code: 437248

Test-taking tip: To respond to questions requiring future predictions, your answers should be based on evidence of past or present behavior and events.

Evaluating Predictions

When making predictions, readers should be able to explain how they developed their prediction. One way readers can defend their thought process is by citing textual evidence. Textual evidence to evaluate reader predictions about literature includes specific synopses of the work, paraphrases of the work or parts of it, and direct quotations from the work. These references to the text must support the prediction by indicating, clearly or unclearly, what will happen later in the story. A text may provide these indications through literary devices such as foreshadowing. Foreshadowing is anything in a text that gives the reader a hint about what is to come by emphasizing the likelihood of an event or development. Foreshadowing can occur through descriptions, exposition, and dialogue. Foreshadowing in dialogue usually occurs when a character gives a warning or expresses a strong feeling that a certain event will occur. Foreshadowing can also occur through irony. However, unlike other forms of foreshadowing, the events that seem the most likely are the opposite of what actually happens. Instances of foreshadowing and irony can be summarized, paraphrased, or quoted to defend a reader's prediction.

Review Video: Textual Evidence for Predictions
Visit mometrix.com/academy and enter code: 261070

Making Inferences and Drawing Conclusions

Inferences are logical conclusions that readers make based on their observations and previous knowledge. An inference is based on both what is found in a passage or a story and what is known from personal experience. For instance, a story may say that a character is frightened and can hear howling in the distance. Based on both what is in the text and personal knowledge, it is a logical conclusion that the character is frightened because he hears the sound of wolves. A good inference is supported by the information in a passage.

Implicit and Explicit Information

By inferring, readers construct meanings from text that are personally relevant. By combining their own schemas or concepts and their background information pertinent to the text with what they read, readers interpret it according to both what the author has conveyed and their own unique

Copyright © Mometrix Media. You have been licensed one copy of this document for personal use only. Any other reproduction or redistribution is strictly prohibited. All rights reserved.
This content is provided for test preparation purposes only and does not imply an endorsement by Mometrix of any particular political, scientific, or religious point of view.

perspectives. Inferences are different from **explicit information**, which is clearly stated in a passage. Authors do not always explicitly spell out every meaning in what they write; many meanings are implicit. Through inference, readers can comprehend implied meanings in the text, and also derive personal significance from it, making the text meaningful and memorable to them. Inference is a natural process in everyday life. When readers infer, they can draw conclusions about what the author is saying, predict what may reasonably follow, amend these predictions as they continue to read, interpret the import of themes, and analyze the characters' feelings and motivations through their actions.

Example of Drawing Conclusions from Inferences

Read the excerpt and decide why Jana finally relaxed.

> Jana loved her job, but the work was very demanding. She had trouble relaxing. She called a friend, but she still thought about work. She ordered a pizza, but eating it did not help. Then, her kitten jumped on her lap and began to purr. Jana leaned back and began to hum a little tune. She felt better.

You can draw the conclusion that Jana relaxed because her kitten jumped on her lap. The kitten purred, and Jana leaned back and hummed a tune. Then she felt better. The excerpt does not explicitly say that this is the reason why she was able to relax. The text leaves the matter unclear, but the reader can infer or make a "best guess" that this is the reason she is relaxing. This is a logical conclusion based on the information in the passage. It is the best conclusion a reader can make based on the information he or she has read. Inferences are based on the information in a passage, but they are not directly stated in the passage.

Test-taking tip: While being tested on your ability to make correct inferences, you must look for **contextual clues**. An answer can be true, but not the best or most correct answer. The contextual clues will help you find the answer that is the **best answer** out of the given choices. Be careful in your reading to understand the context in which a phrase is stated. When asked for the implied meaning of a statement made in the passage, you should immediately locate the statement and read the **context** in which the statement was made. Also, look for an answer choice that has a similar phrase to the statement in question.

Review Video: Inference
Visit mometrix.com/academy and enter code: 379203

Review Video: How to Support a Conclusion
Visit mometrix.com/academy and enter code: 281653

Critical Reading Skills

Opinions, Facts, and Fallacies

Critical thinking skills are mastered through understanding various types of writing and the different purposes authors can have for writing different passages. Every author writes for a purpose. When you understand their purpose and how they accomplish their goal, you will be able to analyze their writing and determine whether or not you agree with their conclusions.

Readers must always be aware of the difference between fact and opinion. A **fact** can be subjected to analysis and proven to be true. An **opinion**, on the other hand, is the author's personal thoughts or feelings and may not be altered by research or evidence. If the author writes that the distance

Copyright © Mometrix Media. You have been licensed one copy of this document for personal use only. Any other reproduction or redistribution is strictly prohibited. All rights reserved.
This content is provided for test preparation purposes only and does not imply an endorsement by Mometrix of any particular political, scientific, or religious point of view.

from New York City to Boston is about two hundred miles, then he or she is stating a fact. If the author writes that New York City is too crowded, then he or she is giving an opinion because there is no objective standard for overpopulation. Opinions are often supported by facts. For instance, an author might use a comparison between the population density of New York City and that of other major American cities as evidence of an overcrowded population. An opinion supported by facts tends to be more convincing. On the other hand, when authors support their opinions with other opinions, readers should employ critical thinking and approach the argument with skepticism.

Review Video: Distinguishing Fact and Opinion
Visit mometrix.com/academy and enter code: 870899

Reliable Sources

When you read an argumentative passage, you need to be sure that facts are presented to the reader from **reliable sources**. An opinion is what the author thinks about a given topic. An opinion is not common knowledge or proven by expert sources, instead the information is the personal beliefs and thoughts of the author. To distinguish between fact and opinion, a reader needs to consider the type of source that is presenting information, the information that backs-up a claim, and the author's motivation to have a certain point-of-view on a given topic. For example, if a panel of scientists has conducted multiple studies on the effectiveness of taking a certain vitamin, then the results are more likely to be factual than those of a company that is selling a vitamin and simply claims that taking the vitamin can produce positive effects. The company is motivated to sell their product, and the scientists are using the scientific method to prove a theory. Remember, if you find sentences that contain phrases such as "I think...", then the statement is an opinion.

Biases

In their attempts to persuade, writers often make mistakes in their thought processes and writing choices. These processes and choices are important to understand so you can make an informed decision about the author's credibility. Every author has a point of view, but authors demonstrate a **bias** when they ignore reasonable counterarguments or distort opposing viewpoints. A bias is evident whenever the author's claims are presented in a way that is unfair or inaccurate. Bias can be intentional or unintentional, but readers should be skeptical of the author's argument in either case. Remember that a biased author may still be correct. However, the author will be correct in spite of, not because of, his or her bias.

A **stereotype** is a bias applied specifically to a group of people or a place. Stereotyping is considered to be particularly abhorrent because it promotes negative, misleading generalizations about people. Readers should be very cautious of authors who use stereotypes in their writing. These faulty assumptions typically reveal the author's ignorance and lack of curiosity.

Review Video: Bias and Stereotype
Visit mometrix.com/academy and enter code: 644829

Persuasion and Rhetoric

Persuasive Techniques

To **appeal using reason**, writers present logical arguments, such as using "If... then... because" statements. To **appeal to emotions**, authors may ask readers how they would feel about something or to put themselves in another's place, present their argument as one that will make the audience feel good, or tell readers how they should feel. To **appeal to character**, **morality**, or **ethics**, authors present their points to readers as the right or most moral choices. Authors cite expert opinions to

Copyright © Mometrix Media. You have been licensed one copy of this document for personal use only. Any other reproduction or redistribution is strictly prohibited. All rights reserved.
This content is provided for test preparation purposes only and does not imply an endorsement by Mometrix of any particular political, scientific, or religious point of view.

show readers that someone very knowledgeable about the subject or viewpoint agrees with the author's claims. **Testimonials**, usually via anecdotes or quotations regarding the author's subject, help build the audience's trust in an author's message through positive support from ordinary people. **Bandwagon appeals** claim that everybody else agrees with the author's argument and persuade readers to conform and agree, also. Authors **appeal to greed** by presenting their choice as cheaper, free, or more valuable for less cost. They **appeal to laziness** by presenting their views as more convenient, easy, or relaxing. Authors also anticipate potential objections and argue against them before audiences think of them, thereby depicting those objections as weak.

Authors can use **comparisons** like analogies, similes, and metaphors to persuade audiences. For example, a writer might represent excessive expenses as "hemorrhaging" money, which the author's recommended solution will stop. Authors can use negative word connotations to make some choices unappealing to readers, and positive word connotations to make others more appealing. Using **humor** can relax readers and garner their agreement. However, writers must take care: ridiculing opponents can be a successful strategy for appealing to readers who already agree with the author, but can backfire by angering other readers. **Rhetorical questions** need no answer, but create effect that can force agreement, such as asking the question, "Wouldn't you rather be paid more than less?" **Generalizations** persuade readers by being impossible to disagree with. Writers can easily make generalizations that appear to support their viewpoints, like saying, "We all want peace, not war" regarding more specific political arguments. **Transfer** and **association** persuade by example: if advertisements show attractive actors enjoying their products, audiences imagine they will experience the same. **Repetition** can also sometimes effectively persuade audiences.

Review Video: Using Rhetorical Strategies for Persuasion
Visit mometrix.com/academy and enter code: 302658

Classical Author Appeals

In his *On Rhetoric,* ancient Greek philosopher Aristotle defined three basic types of appeal used in writing, which he called *pathos, ethos,* and *logos.* ***Pathos*** means suffering or experience and refers to appeals to the emotions (the English word *pathetic* comes from this root). Writing that is meant to entertain audiences, by making them either happy, as with comedy, or sad, as with tragedy, uses *pathos.* Aristotle's *Poetics* states that evoking the emotions of terror and pity is one of the criteria for writing tragedy. ***Ethos*** means character and connotes ideology (the English word *ethics* comes from this root). Writing that appeals to credibility, based on academic, professional, or personal merit, uses *ethos.* ***Logos*** means "I say" and refers to a plea, opinion, expectation, word or speech, account, opinion, or reason (the English word *logic* comes from this root.) Aristotle used it to mean persuasion that appeals to the audience through reasoning and logic to influence their opinions.

Rhetorical Devices

- An **anecdote** is a brief story authors may relate to their argument, which can illustrate their points in a more real and relatable way.
- **Aphorisms** concisely state common beliefs and may rhyme. For example, Benjamin Franklin's "Early to bed and early to rise / Makes a man healthy, wealthy, and wise" is an aphorism.
- **Allusions** refer to literary or historical figures to impart symbolism to a thing or person and to create reader resonance. In John Steinbeck's *Of Mice and Men,* protagonist George's last name is Milton. This alludes to John Milton, who wrote *Paradise Lost,* and symbolizes George's eventual loss of his dream.

Copyright © Mometrix Media. You have been licensed one copy of this document for personal use only. Any other reproduction or redistribution is strictly prohibited. All rights reserved.
This content is provided for test preparation purposes only and does not imply an endorsement by Mometrix of any particular political, scientific, or religious point of view.

- **Satire** exaggerates, ridicules, or pokes fun at human flaws or ideas, as in the works of Jonathan Swift and Mark Twain.
- A **parody** is a form of satire that imitates another work to ridicule its topic or style.
- A **paradox** is a statement that is true despite appearing contradictory.
- **Hyperbole** is overstatement using exaggerated language.
- An **oxymoron** combines seeming contradictions, such as "deafening silence."
- **Analogies** compare two things that share common elements.
- **Similes** (stated comparisons using the words *like* or *as*) and **metaphors** (stated comparisons that do not use *like* or *as*) are considered forms of analogy.
- When using logic to reason with audiences, **syllogism** refers either to deductive reasoning or a deceptive, very sophisticated, or subtle argument.
- **Deductive reasoning** moves from general to specific, **inductive reasoning** from specific to general.
- **Diction** is author word choice that establishes tone and effect.
- **Understatement** achieves effects like contrast or irony by downplaying or describing something more subtly than warranted.
- **Chiasmus** uses parallel clauses, the second reversing the order of the first. Examples include T. S. Eliot's "Has the Church failed mankind, or has mankind failed the Church?" and John F. Kennedy's "Ask not what your country can do for you; ask what you can do for your country."
- **Anaphora** regularly repeats a word or phrase at the beginnings of consecutive clauses or phrases to add emphasis to an idea. A classic example of anaphora was Winston Churchill's emphasis of determination: "[W]e shall fight on the beaches, we shall fight on the landing grounds, we shall fight in the fields and in the streets, we shall fight in the hills; we shall never surrender..."

Reading Informational Texts

Language Use

Literal and Figurative Language

As in fictional literature, informational text also uses both **literal language**, which means just what it says, and **figurative language**, which imparts more than literal meaning. For example, an informational text author might use a simile or direct comparison, such as writing that a racehorse "ran like the wind." Informational text authors also use metaphors or implied comparisons, such as "the cloud of the Great Depression." Imagery may also appear in informational texts to increase the reader's understanding of ideas and concepts discussed in the text.

Explicit and Implicit Information

When informational text states something explicitly, the reader is told by the author exactly what is meant, which can include the author's interpretation or perspective of events. For example, a professor writes, "I have seen students go into an absolute panic just because they weren't able to complete the exam in the time they were allotted." This explicitly tells the reader that the students were afraid, and by using the words "just because," the writer indicates their fear was exaggerated out of proportion relative to what happened. However, another professor writes, "I have had students come to me, their faces drained of all color, saying 'We weren't able to finish the exam.'" This is an example of implicit meaning: the second writer did not state explicitly that the students

Copyright © Mometrix Media. You have been licensed one copy of this document for personal use only. Any other reproduction or redistribution is strictly prohibited. All rights reserved.
This content is provided for test preparation purposes only and does not imply an endorsement by Mometrix of any particular political, scientific, or religious point of view.

were panicked. Instead, he wrote a description of their faces being "drained of all color." From this description, the reader can infer that the students were so frightened that their faces paled.

Review Video: Explicit and Implicit Information
Visit mometrix.com/academy and enter code: 735771

Making Inferences About Informational Text

With informational text, reader comprehension depends not only on recalling important statements and details, but also on reader inferences based on examples and details. Readers add information from the text to what they already know to draw inferences about the text. These inferences help the readers to fill in the information that the text does not explicitly state, enabling them to understand the text better. When reading a nonfictional autobiography or biography, for example, the most appropriate inferences might concern the events in the book, the actions of the subject of the autobiography or biography, and the message the author means to convey. When reading a nonfictional expository (informational) text, the reader would best draw inferences about problems and their solutions, and causes and their effects. When reading a nonfictional persuasive text, the reader will want to infer ideas supporting the author's message and intent.

Structures or Organizational Patterns in Informational Texts

Informational text can be **descriptive**, appealing to the five senses and answering the questions what, who, when, where, and why. Another method of structuring informational text is sequence and order. **Chronological** texts relate events in the sequence that they occurred, from start to finish, while how-to texts organize information into a series of instructions in the sequence in which the steps should be followed. **Comparison-contrast** structures of informational text describe various ideas to their readers by pointing out how things or ideas are similar and how they are different. **Cause and effect** structures of informational text describe events that occurred and identify the causes or reasons that those events occurred. **Problem and solution** structures of informational texts introduce and describe problems and offer one or more solutions for each problem described.

Determining an Informational Author's Purpose

Informational authors' purposes are why they write texts. Readers must determine authors' motivations and goals. Readers gain greater insight into a text by considering the author's motivation. This develops critical reading skills. Readers perceive writing as a person's voice, not simply printed words. Uncovering author motivations and purposes empowers readers to know what to expect from the text, read for relevant details, evaluate authors and their work critically, and respond effectively to the motivations and persuasions of the text. The main idea of a text is what the reader is supposed to understand from reading it; the purpose of the text is why the author has written it and what the author wants readers to do with its information. Authors state some purposes clearly, while other purposes may be unstated but equally significant. When stated purposes contradict other parts of a text, the author may have a hidden agenda. Readers can better evaluate a text's effectiveness, whether they agree or disagree with it, and why they agree or disagree through identifying unstated author purposes.

Identifying Author's Point of View or Purpose

In some informational texts, readers find it easy to identify the author's point of view and purpose, such as when the author explicitly states his or her position and reason for writing. But other texts are more difficult, either because of the content or because the authors give neutral or balanced viewpoints. This is particularly true in scientific texts, in which authors may state the purpose of

Copyright © Mometrix Media. You have been licensed one copy of this document for personal use only. Any other reproduction or redistribution is strictly prohibited. All rights reserved.
This content is provided for test preparation purposes only and does not imply an endorsement by Mometrix of any particular political, scientific, or religious point of view.

their research in the report, but never state their point of view except by interpreting evidence or data.

To analyze text and identify point of view or purpose, readers should ask themselves the following four questions:

1. With what main point or idea does this author want to persuade readers to agree?
2. How does this author's word choice affect the way that readers consider this subject?
3. How do this author's choices of examples and facts affect the way that readers consider this subject?
4. What is it that this author wants to accomplish by writing this text?

Review Video: Understanding the Author's Intent
Visit mometrix.com/academy and enter code: 511819

Review Video: Author's Position
Visit mometrix.com/academy and enter code: 827954

Evaluating Arguments Made by Informational Text Writers

When evaluating an informational text, the first step is to identify the argument's conclusion. Then identify the author's premises that support the conclusion. Try to paraphrase premises for clarification and make the conclusion and premises fit. List all premises first, sequentially numbered, then finish with the conclusion. Identify any premises or assumptions not stated by the author but required for the stated premises to support the conclusion. Read word assumptions sympathetically, as the author might. Evaluate whether premises reasonably support the conclusion. For inductive reasoning, the reader should ask if the premises are true, if they support the conclusion, and if so, how strongly. For deductive reasoning, the reader should ask if the argument is valid or invalid. If all premises are true, then the argument is valid unless the conclusion can be false. If it can, then the argument is invalid. An invalid argument can be made valid through alterations such as the addition of needed premises.

Use of Rhetoric in Informational Texts

There are many ways authors can support their claims, arguments, beliefs, ideas, and reasons for writing in informational texts. For example, authors can appeal to readers' sense of **logic** by communicating their reasoning through a carefully sequenced series of logical steps to help "prove" the points made. Authors can appeal to readers' **emotions** by using descriptions and words that evoke feelings of sympathy, sadness, anger, righteous indignation, hope, happiness, or any other emotion to reinforce what they express and share with their audience. Authors may appeal to the **moral** or **ethical values** of readers by using words and descriptions that can convince readers that something is right or wrong. By relating personal anecdotes, authors can supply readers with more accessible, realistic examples of points they make, as well as appealing to their emotions. They can provide supporting evidence by reporting case studies. They can also illustrate their points by making analogies to which readers can better relate.

Technical Language

Technical Language

Technical language is more impersonal than literary and vernacular language. Passive voice makes the tone impersonal. For example, instead of writing, "We found this a central component of protein metabolism," scientists write, "This was found a central component of protein metabolism." While

Copyright © Mometrix Media. You have been licensed one copy of this document for personal use only. Any other reproduction or redistribution is strictly prohibited. All rights reserved.
This content is provided for test preparation purposes only and does not imply an endorsement by Mometrix of any particular political, scientific, or religious point of view.

science professors have traditionally instructed students to avoid active voice because it leads to first-person ("I" and "we") usage, science editors today find passive voice dull and weak. Many journal articles combine both. Tone in technical science writing should be detached, concise, and professional. While one may normally write, "This chemical has to be available for proteins to be digested," professionals write technically, "The presence of this chemical is required for the enzyme to break the covalent bonds of proteins." The use of technical language appeals to both technical and non-technical audiences by displaying the author or speaker's understanding of the subject and suggesting their credibility regarding the message they are communicating.

TECHNICAL MATERIAL FOR NON-TECHNICAL READERS

Writing about **technical subjects** for **non-technical readers** differs from writing for colleagues because authors place more importance on delivering a critical message than on imparting the maximum technical content possible. Technical authors also must assume that non-technical audiences do not have the expertise to comprehend extremely scientific or technical messages, concepts, and terminology. They must resist the temptation to impress audiences with their scientific knowledge and expertise and remember that their primary purpose is to communicate a message that non-technical readers will understand, feel, and respond to. Non-technical and technical styles include similarities. Both should formally cite any references or other authors' work utilized in the text. Both must follow intellectual property and copyright regulations. This includes the author's protecting his or her own rights, or a public domain statement, as he or she chooses.

Review Video: Technical Passages
Visit mometrix.com/academy and enter code: 478923

NON-TECHNICAL AUDIENCES

Writers of technical or scientific material may need to write for many non-technical audiences. Some readers have no technical or scientific background, and those who do may not be in the same field as the authors. Government and corporate policymakers and budget managers need technical information they can understand for decision-making. Citizens affected by technology or science are a different audience. Non-governmental organizations can encompass many of the preceding groups. Elementary and secondary school programs also need non-technical language for presenting technical subject matter. Additionally, technical authors will need to use non-technical language when collecting consumer responses to surveys, presenting scientific or para-scientific material to the public, writing about the history of science, and writing about science and technology in developing countries.

USE OF EVERYDAY LANGUAGE

Authors of technical information sometimes must write using non-technical language that readers outside their disciplinary fields can comprehend. They should use not only non-technical terms, but also normal, everyday language to accommodate readers whose native language is different than the language the text is written in. For example, instead of writing that "eustatic changes like thermal expansion are causing hazardous conditions in the littoral zone," an author would do better to write that "a rising sea level is threatening the coast." When technical terms cannot be avoided, authors should also define or explain them using non-technical language. Although authors must cite references and acknowledge their use of others' work, they should avoid the kinds of references or citations that they would use in scientific journals—unless they reinforce author messages. They should not use endnotes, footnotes, or any other complicated referential techniques because non-technical journal publishers usually do not accept them. Including high-resolution illustrations, photos, maps, or satellite images and incorporating multimedia into digital publications will enhance non-technical writing about technical subjects. Technical authors may

Copyright © Mometrix Media. You have been licensed one copy of this document for personal use only. Any other reproduction or redistribution is strictly prohibited. All rights reserved.
This content is provided for test preparation purposes only and does not imply an endorsement by Mometrix of any particular political, scientific, or religious point of view.

publish using non-technical language in e-journals, trade journals, specialty newsletters, and daily newspapers.

Reading Argumentative Writing

Author's Argument in Argumentative Writing

In argumentative writing, the argument is a belief, position, or opinion that the author wants to convince readers to believe as well. For the first step, readers should identify the **issue**. Some issues are controversial, meaning people disagree about them. Gun control, foreign policy, and the death penalty are all controversial issues. The next step is to determine the **author's position** on the issue. That position or viewpoint constitutes the author's argument. Readers should then identify the **author's assumptions**: things he or she accepts, believes, or takes for granted without needing proof. Inaccurate or illogical assumptions produce flawed arguments and can mislead readers. Readers should identify what kinds of **supporting evidence** the author offers, such as research results, personal observations or experiences, case studies, facts, examples, expert testimony and opinions, and comparisons. Readers should decide how relevant this support is to the argument.

Review Video: Argumentative Writing
Visit mometrix.com/academy and enter code: 561544

Evaluating an Author's Argument

The first three reader steps to **evaluate an author's argument** are to identify the **author's assumptions**, identify the **supporting evidence**, and decide **whether the evidence is relevant**. For example, if an author is not an expert on a particular topic, then that author's personal experience or opinion might not be relevant. The fourth step is to assess the **author's objectivity**. For example, consider whether the author introduces clear, understandable supporting evidence and facts to support the argument. The fifth step is evaluating whether the author's **argument is complete**. When authors give sufficient support for their arguments and also anticipate and respond effectively to opposing arguments or objections to their points, their arguments are complete. However, some authors omit information that could detract from their arguments. If instead they stated this information and refuted it, it would strengthen their arguments. The sixth step in evaluating an author's argumentative writing is to assess whether the **argument is valid**. Providing clear, logical reasoning makes an author's argument valid. Readers should ask themselves whether the author's points follow a sequence that makes sense, and whether each point leads to the next. The seventh step is to determine whether the author's **argument is credible**, meaning that it is convincing and believable. Arguments that are not valid are not credible, so step seven depends on step six. Readers should be mindful of their own biases as they evaluate and should not expect authors to conclusively prove their arguments, but rather to provide effective support and reason.

Evaluating an Author's Method of Appeal

To evaluate the effectiveness of an appeal, it is important to consider the author's purpose for writing. Any appeals an author uses in their argument must be relevant to the argument's goal. For example, a writer that argues for the reclassification of Pluto, but primarily uses appeals to emotion, will not have an effective argument. This writer should focus on using appeals to logic and support their argument with provable facts. While most arguments should include appeals to logic, emotion, and credibility, some arguments only call for one or two of these types of appeal. Evidence can support an appeal, but the evidence must be relevant to truly strengthen the appeal's effectiveness. If the writer arguing for Pluto's reclassification uses the reasons for Jupiter's

Copyright © Mometrix Media. You have been licensed one copy of this document for personal use only. Any other reproduction or redistribution is strictly prohibited. All rights reserved.
This content is provided for test preparation purposes only and does not imply an endorsement by Mometrix of any particular political, scientific, or religious point of view.

classification as evidence, their argument would be weak. This information may seem relevant because it is related to the classification of planets. However, this classification is highly dependent on the size of the celestial object, and Jupiter is significantly bigger than Pluto. This use of evidence is illogical and does not support the appeal. Even when appropriate evidence and appeals are used, appeals and arguments lose their effectiveness when they create logical fallacies.

EVIDENCE

The term **text evidence** refers to information that supports a main point or minor points and can help lead the reader to a conclusion about the text's credibility. Information used as text evidence is precise, descriptive, and factual. A main point is often followed by supporting details that provide evidence to back up a claim. For example, a passage may include the claim that winter occurs during opposite months in the Northern and Southern hemispheres. Text evidence for this claim may include examples of countries where winter occurs in opposite months. Stating that the tilt of the Earth as it rotates around the sun causes winter to occur at different times in separate hemispheres is another example of text evidence. Text evidence can come from common knowledge, but it is also valuable to include text evidence from credible, relevant outside sources.

Review Video: Textual Evidence
Visit mometrix.com/academy and enter code: 486236

Evidence that supports the thesis and additional arguments needs to be provided. Most arguments must be supported by facts or statistics. A fact is something that is known with certainty, has been verified by several independent individuals, and can be proven to be true. In addition to facts, examples and illustrations can support an argument by adding an emotional component. With this component, you persuade readers in ways that facts and statistics cannot. The emotional component is effective when used alongside objective information that can be confirmed.

CREDIBILITY

The text used to support an argument can be the argument's downfall if the text is not credible. A text is **credible**, or believable, when its author is knowledgeable and objective, or unbiased. The author's motivations for writing the text play a critical role in determining the credibility of the text and must be evaluated when assessing that credibility. Reports written about the ozone layer by an environmental scientist and a hairdresser will have a different level of credibility.

Review Video: Author Credibility
Visit mometrix.com/academy and enter code: 827257

APPEAL TO EMOTION

Sometimes, authors will appeal to the reader's emotion in an attempt to persuade or to distract the reader from the weakness of the argument. For instance, the author may try to inspire the pity of the reader by delivering a heart-rending story. An author also might use the bandwagon approach, in which he suggests that his opinion is correct because it is held by the majority. Some authors resort to name-calling, in which insults and harsh words are delivered to the opponent in an attempt to distract. In advertising, a common appeal is the celebrity testimonial, in which a famous person endorses a product. Of course, the fact that a famous person likes something should not really mean anything to the reader. These and other emotional appeals are usually evidence of poor reasoning and a weak argument.

Review Video: Emotional Language in Literature
Visit mometrix.com/academy and enter code: 759390

Copyright © Mometrix Media. You have been licensed one copy of this document for personal use only. Any other reproduction or redistribution is strictly prohibited. All rights reserved.
This content is provided for test preparation purposes only and does not imply an endorsement by Mometrix of any particular political, scientific, or religious point of view.

Counter Arguments

When authors give both sides to the argument, they build trust with their readers. As a reader, you should start with an undecided or neutral position. If an author presents only his or her side to the argument, then they are not exhibiting credibility and are weakening their argument.

Building common ground with readers can be effective for persuading neutral, skeptical, or opposed readers. Sharing values with undecided readers can allow people to switch positions without giving up what they feel is important. People who may oppose a position need to feel that they can change their minds without betraying who they are as a person. This appeal to having an open mind can be a powerful tool in arguing a position without antagonizing other views. Objections can be countered on a point-by-point basis or in a summary paragraph. Be mindful of how an author points out flaws in counter arguments. If they are unfair to the other side of the argument, then you should lose trust with the author.

Plot and Story Structure

Plot and Story Structure

The **plot** includes the events that happen in a story and the order in which they are told to the reader. There are several types of plot structures, as stories can be told in many ways. The most common plot structure is the chronological plot, which presents the events to the reader in the same order they occur for the characters in the story. Chronological plots usually have five main parts, the **exposition**, **rising action**, the **climax**, **falling action**, and the **resolution**. This type of plot structure guides the reader through the story's events as the characters experience them and is the easiest structure to understand and identify. While this is the most common plot structure, many stories are nonlinear, which means the plot does not sequence events in the same order the characters experience them. Such stories might include elements like flashbacks that cause the story to be nonlinear.

Review Video: How to Make a Story Map
Visit mometrix.com/academy and enter code: 261719

Exposition

The **exposition** is at the beginning of the story and generally takes place before the rising action begins. The purpose of the exposition is to give the reader context for the story, which the author may do by introducing one or more characters, describing the setting or world, or explaining the events leading up to the point where the story begins. The exposition may still include events that contribute to the plot, but the **rising action** and main conflict of the story are not part of the exposition. Some narratives skip the exposition and begin the story with the beginning of the rising action, which causes the reader to learn the context as the story intensifies.

Copyright © Mometrix Media. You have been licensed one copy of this document for personal use only. Any other reproduction or redistribution is strictly prohibited. All rights reserved.
This content is provided for test preparation purposes only and does not imply an endorsement by Mometrix of any particular political, scientific, or religious point of view.

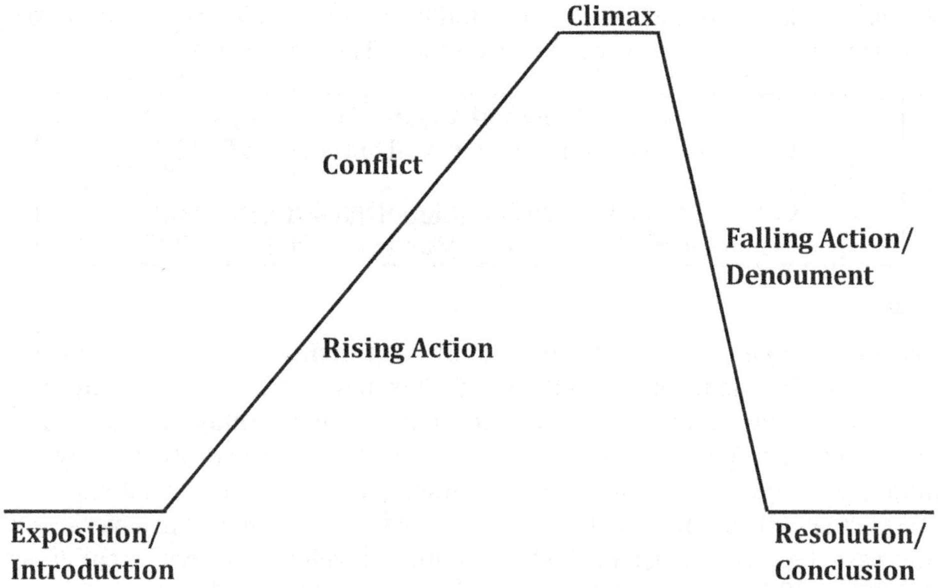

Review Video: Plot Line
Visit mometrix.com/academy and enter code: 944011

CONFLICT

A **conflict** is a problem to be solved. Literary plots typically include one conflict or more. Characters' attempts to resolve conflicts drive the narrative's forward movement. **Conflict resolution** is often the protagonist's primary occupation. Physical conflicts like exploring, wars, and escapes tend to make plots most suspenseful and exciting. Emotional, mental, or moral conflicts tend to make stories more personally gratifying or rewarding for many audiences. Conflicts can be external or internal. A major type of internal conflict is some inner personal battle, or **man versus self**. Major types of external conflicts include **man versus nature**, **man versus man**, and **man versus society**. Readers can identify conflicts in literary plots by identifying the protagonist and antagonist and asking why they conflict, what events develop the conflict, where the climax occurs, and how they identify with the characters.

Read the following paragraph and discuss the type of conflict present:

> Timothy was shocked out of sleep by the appearance of a bear just outside his tent. After panicking for a moment, he remembered some advice he had read in preparation for this trip: he should make noise so the bear would not be startled. As Timothy started to hum and sing, the bear wandered away.

There are three main types of conflict in literature: **man versus man**, **man versus nature**, and **man versus self**. This paragraph is an example of man versus nature. Timothy is in conflict with the bear. Even though no physical conflict like an attack exists, Timothy is pitted against the bear.

Copyright © Mometrix Media. You have been licensed one copy of this document for personal use only. Any other reproduction or redistribution is strictly prohibited. All rights reserved.
This content is provided for test preparation purposes only and does not imply an endorsement by Mometrix of any particular political, scientific, or religious point of view.

Timothy uses his knowledge to "defeat" the bear and keep himself safe. The solution to the conflict is that Timothy makes noise, the bear wanders away, and Timothy is safe.

Review Video: Conflict
Visit mometrix.com/academy and enter code: 559550

Review Video: Determining Relationships in a Story
Visit mometrix.com/academy and enter code: 929925

Rising Action

The **rising action** is the part of the story where conflict **intensifies**. The rising action begins with an event that prompts the main conflict of the story. This may also be called the **inciting incident**. The main conflict generally occurs between the protagonist and an antagonist, but this is not the only type of conflict that may occur in a narrative. After this event, the protagonist works to resolve the main conflict by preparing for an altercation, pursuing a goal, fleeing an antagonist, or doing some other action that will end the conflict. The rising action is composed of several additional events that increase the story's tension. Most often, other developments will occur alongside the growth of the main conflict, such as character development or the development of minor conflicts. The rising action ends with the **climax**, which is the point of highest tension in the story.

Climax

The **climax** is the event in the narrative that marks the height of the story's conflict or tension. The event that takes place at the story's climax will end the rising action and bring about the results of the main conflict. If the conflict was between a good protagonist and an evil antagonist, the climax may be a final battle between the two characters. If the conflict is an adventurer looking for heavily guarded treasure, the climax may be the adventurer's encounter with the final obstacle that protects the treasure. The climax may be made of multiple scenes, but can usually be summarized as one event. Once the conflict and climax are complete, the **falling action** begins.

Falling Action

The **falling action** shows what happens in the story between the climax and the resolution. The falling action often composes a much smaller portion of the story than the rising action does. While the climax includes the end of the main conflict, the falling action may show the results of any minor conflicts in the story. For example, if the protagonist encountered a troll on the way to find some treasure, and the troll demanded the protagonist share the treasure after retrieving it, the falling action would include the protagonist returning to share the treasure with the troll. Similarly, any unexplained major events are usually made clear during the falling action. Once all significant elements of the story are resolved or addressed, the story's resolution will occur. The **resolution** is the end of the story, which shows the final result of the plot's events and shows what life is like for the main characters once they are no longer experiencing the story's conflicts.

Resolution

The way the conflict is **resolved** depends on the type of conflict. The plot of any book starts with the lead up to the conflict, then the conflict itself, and finally the solution, or **resolution**, to the conflict. In **man versus man** conflicts, the conflict is often resolved by two parties coming to some sort of agreement or by one party triumphing over the other party. In **man versus nature** conflicts, the conflict is often resolved by man coming to some realization about some aspect of nature. In

Copyright © Mometrix Media. You have been licensed one copy of this document for personal use only. Any other reproduction or redistribution is strictly prohibited. All rights reserved.
This content is provided for test preparation purposes only and does not imply an endorsement by Mometrix of any particular political, scientific, or religious point of view.

man versus self conflicts, the conflict is often resolved by the character growing or coming to an understanding about part of himself.

THEME

A **theme** is a central idea demonstrated by a passage. Often, a theme is a lesson or moral contained in the text, but it does not have to be. It also is a unifying idea that is used throughout the text; it can take the form of a common setting, idea, symbol, design, or recurring event. A passage can have two or more themes that convey its overall idea. The theme or themes of a passage are often based on **universal themes**. They can frequently be expressed using well-known sayings about life, society, or human nature, such as "Hard work pays off" or "Good triumphs over evil." Themes are not usually stated **explicitly**. The reader must figure them out by carefully reading the passage. Themes are created through descriptive language or events in the plot. The events of a story help shape the themes of a passage.

EXAMPLE

Explain why "if you care about something, you need to take care of it" accurately describes the theme of the following excerpt.

> Luca collected baseball cards, but he wasn't very careful with them. He left them around the house. His dog liked to chew. One day, Luca and his friend Bart were looking at his collection. Then they went outside. When Luca got home, he saw his dog chewing on his cards. They were ruined.

This excerpt tells the story of a boy who is careless with his baseball cards and leaves them lying around. His dog ends up chewing them and ruining them. The lesson is that if you care about something, you need to take care of it. This is the theme, or point, of the story. Some stories have more than one theme, but this is not really true of this excerpt. The reader needs to figure out the theme based on what happens in the story. Sometimes, as in the case of fables, the theme is stated directly in the text. However, this is not usually the case.

Review Video: Themes in Literature
Visit mometrix.com/academy and enter code: 732074

Narrator's Point of View

POINT OF VIEW

Another element that impacts a text is the author's point of view. The **point of view** of a text is the perspective from which a passage is told. An author will always have a point of view about a story before he or she draws up a plot line. The author will know what events they want to take place, how they want the characters to interact, and how they want the story to resolve. An author will also have an opinion on the topic or series of events which is presented in the story that is based on their prior experience and beliefs.

The two main points of view that authors use, especially in a work of fiction, are first person and third person. If the narrator of the story is also the main character, or *protagonist*, the text is written in first-person point of view. In first person, the author writes from the perspective of *I*. Third-person point of view is probably the most common that authors use in their passages. Using third

Copyright © Mometrix Media. You have been licensed one copy of this document for personal use only. Any other reproduction or redistribution is strictly prohibited. All rights reserved.
This content is provided for test preparation purposes only and does not imply an endorsement by Mometrix of any particular political, scientific, or religious point of view.

person, authors refer to each character by using *he* or *she.* In third-person omniscient, the narrator is not a character in the story and tells the story of all of the characters at the same time.

Review Video: Point of View
Visit mometrix.com/academy and enter code: 383336

First-Person Narration

First-person narratives let narrators express inner feelings and thoughts, especially when the narrator is the protagonist as Lemuel Gulliver is in Jonathan Swift's *Gulliver's Travels.* The narrator may be a close friend of the protagonist, like Dr. Watson in Sir Arthur Conan Doyle's *Sherlock Holmes.* Or, the narrator can be less involved with the main characters and plot, like Nick Carraway in F. Scott Fitzgerald's *The Great Gatsby.* When a narrator reports others' narratives, she or he is a "**frame narrator**," like the nameless narrator of Joseph Conrad's *Heart of Darkness* or Mr. Lockwood in Emily Brontë's *Wuthering Heights.* **First-person plural** is unusual but can be effective. Isaac Asimov's *I, Robot*, William Faulkner's *A Rose for Emily*, Maxim Gorky's *Twenty-Six Men and a Girl*, and Jeffrey Eugenides' *The Virgin Suicides* all use first-person plural narration. Author Kurt Vonnegut is the first-person narrator in his semi-autobiographical novel *Timequake.* Also unusual, but effective, is a **first-person omniscient** (rather than the more common third-person omniscient) narrator, like Death in Markus Zusak's *The Book Thief* and the ghost in Alice Sebold's *The Lovely Bones.*

Second-Person Narration

While **second-person** address is very commonplace in popular song lyrics, it is the least used form of narrative voice in literary works. Popular serial books of the 1980s like *Fighting Fantasy* or *Choose Your Own Adventure* employed second-person narratives. In some cases, a narrative combines both second-person and first-person voices, using the pronouns *you* and *I.* This can draw readers into the story, and it can also enable the authors to compare directly "your" and "my" feelings, thoughts, and actions. When the narrator is also a character in the story, as in Edgar Allan Poe's short story "The Tell-Tale Heart" or Jay McInerney's novel *Bright Lights, Big City,* the narrative is better defined as first-person despite it also addressing "you."

Third-Person Narration

Narration in the third person is the most prevalent type, as it allows authors the most flexibility. It is so common that readers simply assume without needing to be informed that the narrator is not a character in the story, or involved in its events. **Third-person singular** is used more frequently than **third-person plural**, though some authors have also effectively used plural. However, both singular and plural are most often included in stories according to which characters are being described. The third-person narrator may be either objective or subjective, and either omniscient or limited. **Objective third-person** narration does not include what the characters described are thinking or feeling, while **subjective third-person** narration does. The **third-person omniscient** narrator knows everything about all characters, including their thoughts and emotions, and all related places, times, and events. However, the **third-person limited** narrator may know everything about a particular character, but is limited to that character. In other words, the narrator cannot speak about anything that character does not know.

Alternating-Person Narration

Although authors more commonly write stories from one point of view, there are also instances wherein they alternate the narrative voice within the same book. For example, they may sometimes use an omniscient third-person narrator and a more intimate first-person narrator at other times. In J. K. Rowling's series of *Harry Potter* novels, she often writes in a third-person limited narrative,

Copyright © Mometrix Media. You have been licensed one copy of this document for personal use only. Any other reproduction or redistribution is strictly prohibited. All rights reserved.
This content is provided for test preparation purposes only and does not imply an endorsement by Mometrix of any particular political, scientific, or religious point of view.

but sometimes changes to narration by characters other than the protagonist. George R. R. Martin's series *A Song of Ice and Fire* changes the point of view to coincide with divisions between chapters. The same technique is used by Erin Hunter (a pseudonym for several authors of the *Warriors, Seekers,* and *Survivors* book series). Authors using first-person narrative sometimes switch to third-person to describe significant action scenes, especially those where the narrator was absent or uninvolved, as Barbara Kingsolver does in her novel *The Poisonwood Bible.*

Setting, Mood, and Tone

Setting and Time Frame

A literary text has both a setting and time frame. A **setting** is the place in which the story as a whole is set. The **time frame** is the period in which the story is set. This may refer to the historical period the story takes place in or if the story takes place over a single day. Both setting and time frame are relevant to a text's meaning because they help the reader place the story in time and space. An author uses setting and time frame to anchor a text, create a mood, and enhance its meaning. This helps a reader understand why a character acts the way he does, or why certain events in the story are important. The setting impacts the **plot** and character **motivations**, while the time frame helps place the story in **chronological context**.

Example

Read the following excerpt from The Adventures of Huckleberry Finn by Mark Twain and analyze the relevance of setting to the text's meaning:

> We said there warn't no home like a raft, after all. Other places do seem so cramped up and smothery, but a raft don't. You feel mighty free and easy and comfortable on a raft.

This excerpt from *The Adventures of Huckleberry Finn* by Mark Twain reveals information about the **setting** of the book. By understanding that the main character, Huckleberry Finn, lives on a raft, the reader can place the story on a river, in this case, the Mississippi River in the South before the Civil War. The information about the setting also gives the reader clues about the **character** of Huck Finn: he clearly values independence and freedom, and he likes the outdoors. The information about the setting in the quote helps the reader to better understand the rest of the text.

Syntax and Word Choice

Authors use words and **syntax**, or sentence structure, to make their texts unique, convey their own writing style, and sometimes to make a point or emphasis. They know that word choice and syntax contribute to the reader's understanding of the text as well as to the tone and mood of a text.

Review Video: Syntax
Visit mometrix.com/academy and enter code: 242280

Mood and Tone

Mood is a story's atmosphere, or the feelings the reader gets from reading it. The way authors set the mood in writing is comparable to the way filmmakers use music to set the mood in movies. Instead of music, though, writers judiciously select descriptive words to evoke certain **moods**. The mood of a work may convey joy, anger, bitterness, hope, gloom, fear, apprehension, or any other emotion the author wants the reader to feel. In addition to vocabulary choices, authors also use figurative expressions, particular sentence structures, and choices of diction that project and reinforce the moods they want to create. Whereas mood is the reader's emotions evoked by reading

Copyright © Mometrix Media. You have been licensed one copy of this document for personal use only. Any other reproduction or redistribution is strictly prohibited. All rights reserved.
This content is provided for test preparation purposes only and does not imply an endorsement by Mometrix of any particular political, scientific, or religious point of view.

what is written, **tone** is the emotions and attitudes of the writer that she or he expresses in the writing. Authors use the same literary techniques to establish tone as they do to establish mood. An author may use a humorous tone, an angry or sad tone, a sentimental or unsentimental tone, or something else entirely.

Mood and Tone in The Great Gatsby

To understand the difference between mood and tone, look at this excerpt from F. Scott Fitzgerald's *The Great Gatsby*. In this passage, Nick Caraway, the novel's narrator, is describing his affordable house, which sits in a neighborhood full of expensive mansions.

> "I lived at West Egg, the—well the less fashionable of the two, though this is a most superficial tag to express the bizarre and not a little sinister contrast between them. My house was at the very tip of the egg, only fifty yard from the Sound, and squeezed between two huge places that rented for twelve or fifteen thousand a season ... My own house was an eyesore, but it was a small eyesore, and it had been overlooked, so I had a view of the water, a partial view of my neighbor's lawn, and the consoling proximity of millionaires—all for eighty dollars a month."

In this description, the mood created for the reader does not match the tone created through the narrator. The mood in this passage is one of dissatisfaction and inferiority. Nick compares his home to his neighbors', saying he lives in the "less fashionable" neighborhood and that his house is "overlooked," an "eyesore," and "squeezed between two huge" mansions. He also adds that his placement allows him the "consoling proximity of millionaires." A literal reading of these details leads the reader to have negative feelings toward Nick's house and his economic inferiority to his neighbors, creating the mood.

However, Fitzgerald also conveys an opposing attitude, or tone, through Nick's description. Nick calls the distinction between the neighborhoods "superficial," showing a suspicion of the value suggested by the neighborhoods' titles, properties, and residents. Nick also undermines his critique of his own home by calling it "a small eyesore" and claiming it has "been overlooked." However, he follows these statements with a description of his surroundings, claiming that he has "a view of the water" and can see some of his wealthy neighbor's property from his home, and a comparison between the properties' rent. While the mental image created for the reader depicts a small house shoved between looming mansions, the tone suggests that Nick enjoys these qualities about his home, or at least finds it charming. He acknowledges its shortcomings, but includes the benefits of his home's unassuming appearance.

Review Video: Style, Tone, and Mood
Visit mometrix.com/academy and enter code: 416961

Historical and Social Context

Fiction that is heavily influenced by a historical or social context cannot be comprehended as the author intended if the reader does not keep this context in mind. Many important elements of the text will be influenced by any context, including symbols, allusions, settings, and plot events. These contexts, as well as the identity of the work's author, can help to inform the reader about the author's concerns and intended meanings. For example, George Orwell published his novel *1984* in the year 1949, soon after the end of World War II. At that time, following the defeat of the Nazis, the Cold War began between the Western Allied nations and the Eastern Soviet Communists. People were therefore concerned about the conflict between the freedoms afforded by Western democracies versus the oppression represented by Communism. Orwell had also previously fought

Copyright © Mometrix Media. You have been licensed one copy of this document for personal use only. Any other reproduction or redistribution is strictly prohibited. All rights reserved.
This content is provided for test preparation purposes only and does not imply an endorsement by Mometrix of any particular political, scientific, or religious point of view.

in the Spanish Civil War against a Spanish regime that he and his fellows viewed as oppressive. From this information, readers can infer that Orwell was concerned about oppression by totalitarian governments. This informs *1984*'s story of Winston Smith's rebellion against the oppressive "Big Brother" government, of the fictional dictatorial state of Oceania, and his capture, torture, and ultimate conversion by that government. Some literary theories also seek to use historical and social contexts to reveal deeper meanings and implications in a text.

Figurative Language

Literal and Figurative Meaning

When language is used **literally**, the words mean exactly what they say and nothing more. When language is used **figuratively**, the words mean something beyond their literal meaning. For example, "The weeping willow tree has long, trailing branches and leaves" is a literal description. But "The weeping willow tree looks as if it is bending over and crying" is a figurative description—specifically, a **simile** or stated comparison. Another figurative language form is **metaphor**, or an implied comparison. A good example is the metaphor of a city, state, or city-state as a ship, and its governance as sailing that ship. Ancient Greek lyrical poet Alcaeus is credited with first using this metaphor, and ancient Greek tragedian Aeschylus then used it in *Seven Against Thebes,* and then Plato used it in the *Republic.*

Figures of Speech

A **figure of speech** is a verbal expression whose meaning is figurative rather than literal. For example, the phrase "butterflies in the stomach" does not refer to actual butterflies in a person's stomach. It is a metaphor representing the fluttery feelings experienced when a person is nervous or excited—or when one "falls in love," which does not mean physically falling. "Hitting a sales target" does not mean physically hitting a target with arrows as in archery; it is a metaphor for meeting a sales quota. "Climbing the ladder of success" metaphorically likens advancing in one's career to ascending ladder rungs. Similes, such as "light as a feather" (meaning very light, not a feather's actual weight), and hyperbole, like "I'm starving/freezing/roasting," are also figures of speech. Figures of speech are often used and crafted for emphasis, freshness of expression, or clarity.

Review Video: Figures of Speech
Visit mometrix.com/academy and enter code: 111295

Figurative Language

Figurative language extends past the literal meanings of words. It offers readers new insight into the people, things, events, and subjects covered in a work of literature. Figurative language also enables readers to feel they are sharing the authors' experiences. It can stimulate the reader's senses, make comparisons that readers find intriguing or even startling, and enable readers to view the world in different ways. When looking for figurative language, it is important to consider the context of the sentence or situation. Phrases that appear out of place or make little sense when read literally are likely instances of figurative language. Once figurative language has been recognized, context is also important to determining the type of figurative language being used and its function. For example, when a comparison is being made, a metaphor or simile is likely being used. This means the comparison may emphasize or create irony through the things being compared. Seven

Copyright © Mometrix Media. You have been licensed one copy of this document for personal use only. Any other reproduction or redistribution is strictly prohibited. All rights reserved.
This content is provided for test preparation purposes only and does not imply an endorsement by Mometrix of any particular political, scientific, or religious point of view.

specific types of figurative language include: alliteration, onomatopoeia, personification, imagery, similes, metaphors, and hyperbole.

Review Video: Figurative Language
Visit mometrix.com/academy and enter code: 584902

Alliteration and Onomatopoeia

Alliteration describes a series of words beginning with the same sounds. **Onomatopoeia** uses words imitating the sounds of things they name or describe. For example, in his poem "Come Down, O Maid," Alfred Tennyson writes of "The moan of doves in immemorial elms, / And murmuring of innumerable bees." The word "moan" sounds like some sounds doves make, "murmuring" represents the sounds of bees buzzing. Onomatopoeia also includes words that are simply meant to represent sounds, such as "meow," "kaboom," and "whoosh."

Review Video: Alliteration in Everyday Expressions
Visit mometrix.com/academy and enter code: 462837

Personification

Another type of figurative language is **personification**. This is describing a non-human thing, like an animal or an object, as if it were human. The general intent of personification is to describe things in a manner that will be comprehensible to readers. When an author states that a tree *groans* in the wind, he or she does not mean that the tree is emitting a low, pained sound from a mouth. Instead, the author means that the tree is making a noise similar to a human groan. Of course, this personification establishes a tone of sadness or suffering. A different tone would be established if the author said that the tree was *swaying* or *dancing*. Alfred Tennyson's poem "The Eagle" uses all of these types of figurative language: "He clasps the crag with crooked hands." Tennyson used alliteration, repeating /k/ and /kr/ sounds. These hard-sounding consonants reinforce the imagery, giving visual and tactile impressions of the eagle.

Review Video: Personification
Visit mometrix.com/academy and enter code: 260066

Similes and Metaphors

Similes are stated comparisons using "like" or "as." Similes can be used to stimulate readers' imaginations and appeal to their senses. Because a simile includes *like* or *as,* the device creates more space between the description and the thing being described than a metaphor does. If an author says that *a house was like a shoebox*, then the tone is different than the author saying that the house *was* a shoebox. Authors will choose between a metaphor and a simile depending on their intended tone.

Similes also help compare fictional characters to well-known objects or experiences, so the reader can better relate to them. William Wordsworth's poem about "Daffodils" begins, "I wandered lonely as a cloud." This simile compares his loneliness to that of a cloud. It is also personification, giving a cloud the human quality loneliness. In his novel *Lord Jim* (1900), Joseph Conrad writes in Chapter 33, "I would have given anything for the power to soothe her frail soul, tormenting itself in its invincible ignorance like a small bird beating about the cruel wires of a cage." Conrad uses the word

Copyright © Mometrix Media. You have been licensed one copy of this document for personal use only. Any other reproduction or redistribution is strictly prohibited. All rights reserved.
This content is provided for test preparation purposes only and does not imply an endorsement by Mometrix of any particular political, scientific, or religious point of view.

"like" to compare the girl's soul to a small bird. His description of the bird beating at the cage shows the similar helplessness of the girl's soul to gain freedom.

Review Video: Similes
Visit mometrix.com/academy and enter code: 642949

A **metaphor** is a type of figurative language in which the writer equates something with another thing that is not particularly similar, instead of using *like* or *as*. For instance, *the bird was an arrow arcing through the sky*. In this sentence, the arrow is serving as a metaphor for the bird. The point of a metaphor is to encourage the reader to consider the item being described in a *different way*. Let's continue with this metaphor for a flying bird. You are asked to envision the bird's flight as being similar to the arc of an arrow. So, you imagine the flight to be swift and bending. Metaphors are a way for the author to describe an item *without being direct and obvious*. This literary device is a lyrical and suggestive way of providing information. Note that the reference for a metaphor will not always be mentioned explicitly by the author. Consider the following description of a forest in winter: *Swaying skeletons reached for the sky and groaned as the wind blew through them.* In this example, the author is using *skeletons* as a metaphor for leafless trees. This metaphor creates a spooky tone while inspiring the reader's imagination.

LITERARY EXAMPLES OF METAPHOR

A **metaphor** is an implied comparison, i.e., it compares something to something else without using "like", "as", or other comparative words. For example, in "The Tyger" (1794), William Blake writes, "Tyger Tyger, burning bright, / In the forests of the night." Blake compares the tiger to a flame not by saying it is like a fire, but by simply describing it as "burning." Henry Wadsworth Longfellow's poem "O Ship of State" (1850) uses an extended metaphor by referring consistently throughout the entire poem to the state, union, or republic as a seagoing vessel, referring to its keel, mast, sail, rope, anchors, and to its braving waves, rocks, gale, tempest, and "false lights on the shore." Within the extended metaphor, Wordsworth uses a specific metaphor: "the anchors of thy hope!"

TED HUGHES' ANIMAL METAPHORS

Ted Hughes frequently used animal metaphors in his poetry. In "The Thought Fox," a model of concise, structured beauty, Hughes characterizes the poet's creative process with succinct, striking imagery of an idea entering his head like a wild fox. Repeating "loneliness" in the first two stanzas emphasizes the poet's lonely work: "Something else is alive / Beside the clock's loneliness." He treats an idea's arrival as separate from himself. Three stanzas detail in vivid images a fox's approach from the outside winter forest at starless midnight—its nose, "Cold, delicately" touching twigs and leaves; "neat" paw prints in snow; "bold" body; brilliant green eyes; and self-contained, focused progress—"Till, with a sudden sharp hot stink of fox," he metaphorically depicts poetic inspiration as the fox's physical entry into "the dark hole of the head." Hughes ends by summarizing his vision of a poet as an interior, passive idea recipient, with the outside world unchanged: "The window is starless still; the clock ticks, / The page is printed."

Review Video: Metaphors in Writing
Visit mometrix.com/academy and enter code: 133295

METONYMY

Metonymy is naming one thing with words or phrases of a closely related thing. This is similar to metaphor. However, the comparison has a close connection, unlike metaphor. An example of metonymy is to call the news media *the press*. Of course, *the press* is the machine that prints newspapers. Metonymy is a way of naming something without using the same name constantly.

Copyright © Mometrix Media. You have been licensed one copy of this document for personal use only. Any other reproduction or redistribution is strictly prohibited. All rights reserved.
This content is provided for test preparation purposes only and does not imply an endorsement by Mometrix of any particular political, scientific, or religious point of view.

SYNECDOCHE

Synecdoche points to the whole by naming one of the parts. An example of synecdoche would be calling a construction worker a *hard hat*. Like metonymy, synecdoche is an easy way of naming something without having to overuse a name. The device allows writers to highlight pieces of the thing being described. For example, referring to businessmen as *suits* suggests professionalism and unity.

HYPERBOLE

Hyperbole is excessive exaggeration used for humor or emphasis rather than for literal meaning. For example, in *To Kill a Mockingbird*, Harper Lee wrote, "People moved slowly then. There was no hurry, for there was nowhere to go, nothing to buy and no money to buy it with, nothing to see outside the boundaries of Maycomb County." This was not literally true; Lee exaggerates the scarcity of these things for emphasis. In "Old Times on the Mississippi," Mark Twain wrote, "I... could have hung my hat on my eyes, they stuck out so far." This is not literal, but makes his description vivid and funny. In his poem "As I Walked Out One Evening", W. H. Auden wrote, "I'll love you, dear, I'll love you / Till China and Africa meet, / And the river jumps over the mountain / And the salmon sing in the street." He used things not literally possible to emphasize the duration of his love.

UNDERSTATEMENT

Understatement is the opposite of hyperbole. This device discounts or downplays something. Think about someone who climbs Mount Everest. Then, they say that the journey was *a little stroll*. As with other types of figurative language, understatement has a range of uses. The device may show self-defeat or modesty as in the Mount Everest example. However, some may think of understatement as false modesty (i.e., an attempt to bring attention to you or a situation). For example, a woman is praised on her diamond engagement ring. The woman says, *Oh, this little thing?* Her understatement might be heard as stuck-up or unfeeling.

Review Video: Hyperbole and Understatement
Visit mometrix.com/academy and enter code: 308470

Literary Devices

LITERARY IRONY

In literature, irony demonstrates the opposite of what is said or done. The three types of irony are **verbal irony**, **situational irony**, and **dramatic irony**. Verbal irony uses words opposite to the meaning. Sarcasm may use verbal irony. One common example is describing something that is confusing as "clear as mud." For example, in his 1986 movie *Hannah and Her Sisters,* author, director, and actor Woody Allen says to his character's date, "I had a great evening; it was like the Nuremburg Trials." Notice these employ similes. In situational irony, what happens contrasts with what was expected. O. Henry's short story *The Gift of the Magi* uses situational irony: a husband and wife each sacrifice their most prized possession to buy each other a Christmas present. The irony is that she sells her long hair to buy him a watch fob, while he sells his heirloom pocket-watch to buy her the jeweled combs for her hair she had long wanted; in the end, neither of them can use their gifts. In dramatic irony, narrative informs audiences of more than its characters know. For example,

Copyright © Mometrix Media. You have been licensed one copy of this document for personal use only. Any other reproduction or redistribution is strictly prohibited. All rights reserved.
This content is provided for test preparation purposes only and does not imply an endorsement by Mometrix of any particular political, scientific, or religious point of view.

in *Romeo and Juliet,* the audience is made aware that Juliet is only asleep, while Romeo believes her to be dead, which then leads to Romeo's death.

Review Video: Irony
Visit mometrix.com/academy and enter code: 374204

IDIOMS

Idioms create comparisons, and often take the form of similes or metaphors. Idioms are always phrases and are understood to have a meaning that is different from its individual words' literal meaning. For example, "break a leg" is a common idiom that is used to wish someone luck or tell them to perform well. Literally, the phrase "break a leg" means to injure a person's leg, but the phrase takes on a different meaning when used as an idiom. Another example is "call it a day," which means to temporarily stop working on a task, or find a stopping point, rather than literally referring to something as "a day." Many idioms are associated with a region or group. For example, an idiom commonly used in the American South is "'til the cows come home." This phrase is often used to indicate that something will take or may last for a very long time, but not that it will literally last until the cows return to where they reside.

ALLUSION

An allusion is an uncited but recognizable reference to something else. Authors use language to make allusions to places, events, artwork, and other books in order to make their own text richer. For example, an author may allude to a very important text in order to make his own text seem more important. Martin Luther King, Jr. started his "I Have a Dream" speech by saying "Five score years ago..." This is a clear allusion to President Abraham Lincoln's "Gettysburg Address" and served to remind people of the significance of the event. An author may allude to a place to ground his text or make a cultural reference to make readers feel included. There are many reasons that authors make allusions.

Review Video: Allusions
Visit mometrix.com/academy and enter code: 294065

COMIC RELIEF

Comic relief is the use of comedy by an author to break up a dramatic or tragic scene and infuse it with a bit of **lightheartedness**. In William Shakespeare's *Hamlet,* two gravediggers digging the grave for Ophelia share a joke while they work. The death and burial of Ophelia are tragic moments that directly follow each other. Shakespeare uses an instance of comedy to break up the tragedy and give his audience a bit of a break from the tragic drama. Authors sometimes use comic relief so that their work will be less depressing; other times they use it to create irony or contrast between the darkness of the situation and the lightness of the joke. Often, authors will use comedy to parallel what is happening in the tragic scenes.

Review Video: Comic Relief
Visit mometrix.com/academy and enter code: 779604

FORESHADOWING

Foreshadowing is a device authors use to give readers **hints** about events that will take place later in a story. Foreshadowing most often takes place through a character's dialogue or actions. Sometimes the character will know what is going to happen and will purposefully allude to future events. For example, consider a protagonist who is about to embark on a journey through the woods. Just before the protagonist begins the trip, another character says, "Be careful, you never

Copyright © Mometrix Media. You have been licensed one copy of this document for personal use only. Any other reproduction or redistribution is strictly prohibited. All rights reserved.
This content is provided for test preparation purposes only and does not imply an endorsement by Mometrix of any particular political, scientific, or religious point of view.

know what could be out in those woods!" This alerts the reader that the woods may be dangerous and prompts the reader to expect something to attack the protagonist in the woods. This is an example of foreshadowing through warning. Alternatively, a character may unknowingly foreshadow later events. For example, consider a story where a brother and sister run through their house and knock over a vase and break it. The brother says, "Don't worry, we'll clean it up! Mom will never know!" However, the reader knows that their mother will most likely find out what they have done, so the reader expects the siblings to later get in trouble for running, breaking the vase, and hiding it from their mother.

SYMBOLISM

Symbolism describes an author's use of a **symbol**, an element of the story that **represents** something else. Symbols can impact stories in many ways, including deepening the meaning of a story or its elements, comparing a story to another work, or foreshadowing later events in a story. Symbols can be objects, characters, colors, numbers, or anything else the author establishes as a symbol. Symbols can be clearly established through direct comparison or repetition, but they can also be established subtly or gradually over a large portion of the story. Another form of symbolism is **allusion**, which is when something in a story is used to prompt the reader to think about another work. Many well-known works use **Biblical allusions**, which are allusions to events or details in the Bible that inform a work or an element within it.

Chapter Quiz

Ready to see how well you retained what you just read? Scan the QR code to go directly to the chapter quiz interface for this study guide. If you're using a computer, simply visit the bonus page at **mometrix.com/bonus948/ibewapt** and click the Chapter Quizzes link.

Copyright © Mometrix Media. You have been licensed one copy of this document for personal use only. Any other reproduction or redistribution is strictly prohibited. All rights reserved.
This content is provided for test preparation purposes only and does not imply an endorsement by Mometrix of any particular political, scientific, or religious point of view.

IBEW Practice Test #1

Want to take this practice test in an online interactive format?
Check out the bonus page, which includes interactive practice questions and much more: **mometrix.com/bonus948/ibewapt**

Algebra and Functions

1. Solve the system of equations.

$$3x + 4y = 2$$
$$2x + 6y = -2$$

a. $\left(0, \frac{1}{2}\right)$
b. $\left(\frac{2}{5}, \frac{1}{5}\right)$
c. $(2, -1)$
d. $\left(-1, \frac{5}{4}\right)$

2. A metal rod used in manufacturing must be as close as possible to 15 inches in length. The tolerance of the length, L, in inches, is specified by the inequality $|L - 15| \leq 0.01$. What is the minimum length permissible for the rod?

a. 14.9 inches
b. 14.99 inches
c. 15.01 inches
d. 15.1 inches

3. Which of the following statements is (are) true when $f(x) = \frac{x^2 - x - 2}{x^3 - 7x + 6}$?

I. The graph $f(x)$ has vertical asymptotes at $x = -3$, $x = 1$, and $x = 2$.
II. The x- and y-intercepts of the graph of $f(x)$ are both $-\frac{1}{3}$.

a. I only
b. II only
c. I and II
d. Neither statement is true.

4. What is the intersection of two lines formed by the equations $y = 2x + 3$ and $y = x - 5$?

a. (5,3)
b. (8,13)
c. (−4,13)
d. (−8, −13)

Copyright © Mometrix Media. You have been licensed one copy of this document for personal use only. Any other reproduction or redistribution is strictly prohibited. All rights reserved.
This content is provided for test preparation purposes only and does not imply an endorsement by Mometrix of any particular political, scientific, or religious point of view.

5. What is the average of $\frac{7}{5}$ and 1.4?

a. 1.4
b. 2.8
c. 4.2
d. 7.4

6. Which number equals 2^{-3}?

a. $\frac{1}{2}$
b. $\frac{1}{4}$
c. $\frac{1}{8}$
d. $\frac{1}{16}$

7. Which of these relationships represents y as a function of x?

a. $x^2 + y^2 = 16$

b.

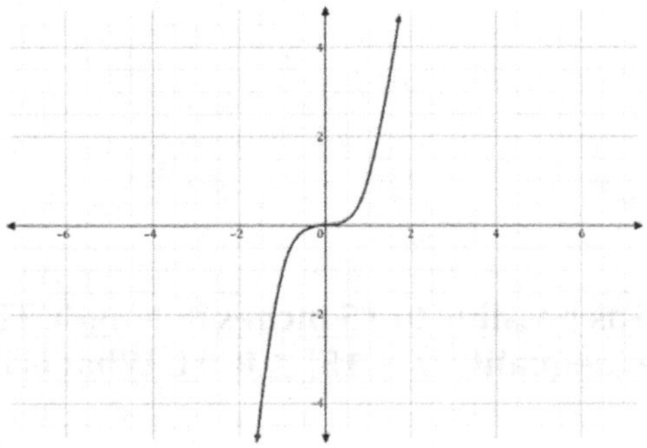

c. $y = \pm\sqrt{x}$

d.

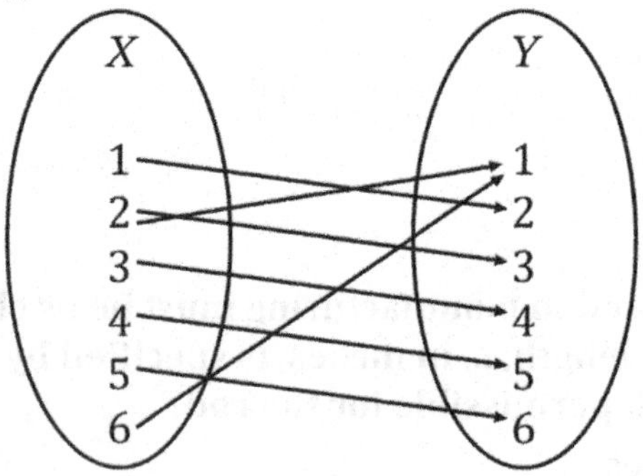

8. What fractional part of an hour is 400 seconds?

a. $\frac{1}{6}$
b. $\frac{1}{7}$
c. $\frac{1}{8}$
d. $\frac{1}{9}$

Copyright © Mometrix Media. You have been licensed one copy of this document for personal use only. Any other reproduction or redistribution is strictly prohibited. All rights reserved.
This content is provided for test preparation purposes only and does not imply an endorsement by Mometrix of any particular political, scientific, or religious point of view.

9. Which of these equations most closely resembles the graph below?

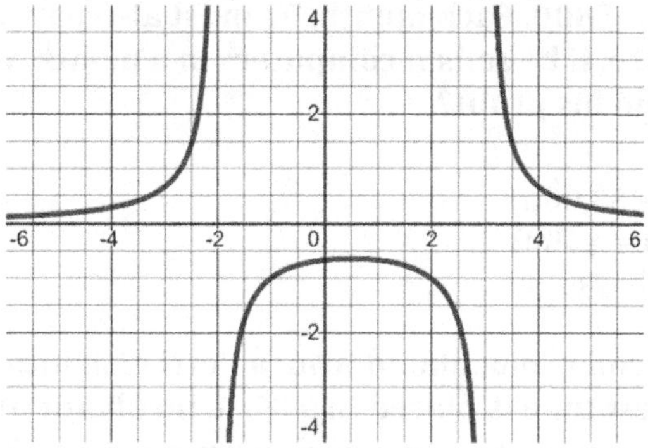

a. $y = \frac{2x+4}{x^2-x-6}$
b. $y = \frac{4}{x^2-x-6}$
c. $y = \frac{2}{x+2} + \frac{2}{x-3}$
d. $y = \frac{x+4}{x^2-x-3}$

10. Which of these statements is NOT necessarily true when $f(x) = \log_b x$ and $b > 1$?

a. The x-intercept of the graph of $f(x)$ is $(1, 0)$.
b. The graph of $f(x)$ passes through $(b, 1)$.
c. $f(x) < 0$ when $0 < x < 1$.
d. If $g(x) = x^b$, the graph of $f(x)$ is symmetric to the graph of $g(x)$ with respect to $y = x$.

11. Mark is leaving a job site and moving equipment to Phoenix which is located 210 miles north. He drives the first ten miles in 12 minutes. If he continues at the same rate, what will the total travel time be?

a. 3 hours 15 minutes
b. 3 hours 45 minutes
c. 4 hours 12 minutes
d. 4 hours 20 minutes

12. Put the following numbers in order from the least to greatest: $2^3, 4^2, 6^0, 9, 10^1$.

a. $2^3, 4^2, 6^0, 9, 10^1$
b. $6^0, 9, 10^1, 2^3, 4^2$
c. $10^1, 2^3, 6^0, 9, 4^2$
d. $6^0, 2^3, 9, 10^1, 4^2$

13. Which of the following is a solution to the inequality $4x - 12 < 4$?

a. 7
b. 6
c. 4
d. 3

Copyright © Mometrix Media. You have been licensed one copy of this document for personal use only. Any other reproduction or redistribution is strictly prohibited. All rights reserved.
This content is provided for test preparation purposes only and does not imply an endorsement by Mometrix of any particular political, scientific, or religious point of view.

14. Rafael has a business selling computers. He buys computers from the manufacturer for \$450 each and sells them for \$800. Each month, he must also pay fixed costs of \$3,000 for rent and utilities for his store. If he sells n computers in a month, which of the following equations can be used to find his profit?

a. $P = n(\$800 - \$450)$
b. $P = n(\$800 - \$450 - \$3{,}000)$
c. $P = \$3{,}000 \times n(\$800 - \$450)$
d. $P = n(\$800 - \$450) - \$3{,}000$

15. A colony of Escherichia coli is inoculated from a Petri dish into a test tube containing 50 mL of nutrient broth. The test tube is placed in a 37 °C incubator and agitator. After one hour, the number of bacteria in the test tube is determined to be 8×10^6. Given that the doubling time of Escherichia coli is 20 minutes with agitation at 37 °C, approximately how many bacteria should the test tube contain after eight hours of growth?

a. 2.56×10^8
b. 2.05×10^9
c. 1.7×10^{14}
d. 1.7×10^{13}

16. Simplify: $|7 - 5| - |5 - 7|$

a. -2
b. 0
c. 2
d. 4

17. Which of these can be modeled by a quadratic function?

a. The path of a sound wave
b. The path of a bullet
c. The distance an object travels over time when the rate is constant
d. Radioactive decay

Copyright © Mometrix Media. You have been licensed one copy of this document for personal use only. Any other reproduction or redistribution is strictly prohibited. All rights reserved.
This content is provided for test preparation purposes only and does not imply an endorsement by Mometrix of any particular political, scientific, or religious point of view.

18. The graph below shows the number of miles Jen runs each day, Monday through Friday. What fraction of the time does she run at least four miles?

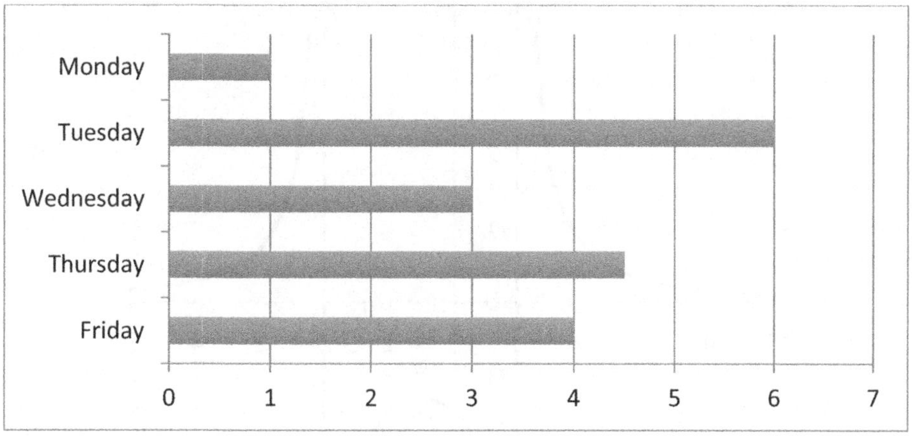

a. $\frac{3}{7}$
b. $\frac{3}{2}$
c. $\frac{2}{5}$
d. $\frac{3}{5}$

19. Solve $4x^3y^2 - 16x^2y = -16x^3y + 4x^2y^2$ for x.

a. $x = \{0\}$
b. $x = \{0,1\}$
c. $x = \{-4\}$
d. $x = \{-4,0,1\}$

20. A bullet travels at 5×10^6 feet per hour. If it strikes its target in 2×10^{-4} hours, how far has it traveled?

a. 50 feet
b. 100 feet
c. 200 feet
d. 1,000 feet

Copyright © Mometrix Media. You have been licensed one copy of this document for personal use only. Any other reproduction or redistribution is strictly prohibited. All rights reserved.
This content is provided for test preparation purposes only and does not imply an endorsement by Mometrix of any particular political, scientific, or religious point of view.

21. Which of these equations is represented by the graph below?

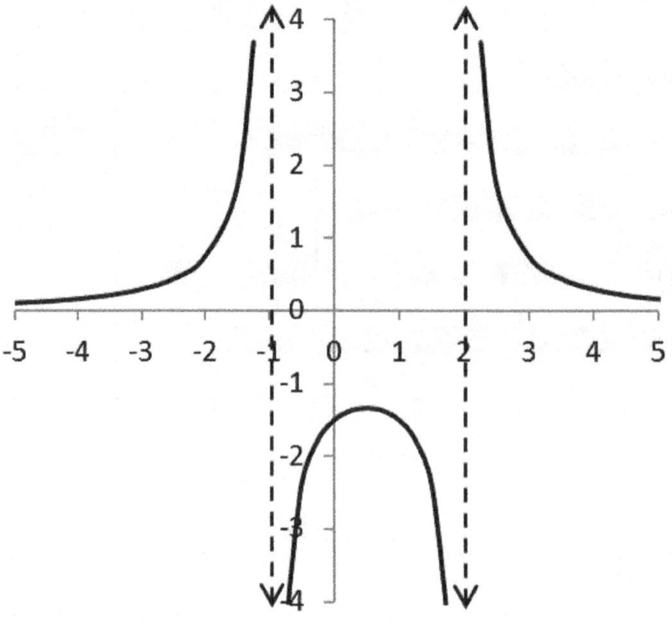

a. $y = \frac{3}{x^2-x-2}$
b. $y = \frac{3x+3}{x^2-x-2}$
c. $y = \frac{1}{x+1} + \frac{1}{x-2}$
d. None of these

22. Which of these statements is NOT necessarily true when $f(x) = \log_b x$ and $b > 1$?

a. The x-intercept of the graph of $f(x)$ is 0.
b. The graph of $f(x)$ passes through $(b, 1)$
c. $f(x) < 0$ when $0 < x < 1$
d. If $g(x) = b^x$, the graph of $f(x)$ is symmetric to the graph of $g(x)$ with respect to $y = x$.

23. Marcus is mowing yards and doing odd jobs to earn money for a new video game system that costs $325. Marcus only charges $6.50 per hour. Which of the following equations represents the number of hours Marcus needs to work to earn $325?

a. $6.50x = 325$
b. $6.50 + x = 325$
c. $325x = 6.50$
d. $6.50x + 325 = x$

Copyright © Mometrix Media. You have been licensed one copy of this document for personal use only. Any other reproduction or redistribution is strictly prohibited. All rights reserved.
This content is provided for test preparation purposes only and does not imply an endorsement by Mometrix of any particular political, scientific, or religious point of view.

24. Given this partial table of values for $f(x)$ and $g(x)$, find $f(g(-3))$. (Assume that $f(x)$ and $g(x)$ are the simplest polynomials that fit the data.)

x	$f(x)$	$g(x)$
-2	12	-6
-1	3	-4
0	0	-2
1	3	0
2	12	2
3		

a. 192
b. 104
c. 64
d. -8

25. If $a = 3$ and $b = -2$, what is the value of $a^2 + 3ab - b^2$?

a. -13
b. -4
c. 5
d. 12

26. Which of these can be modeled by a sinusoidal function?

a. The path of a sound wave
b. The path of a bullet
c. The distance an object travels over time when the rate is constant
d. Radioactive decay

27. The scatterplot below shows Julio's time (in hours) on the y-axis spent practicing his trumpet on the days (x-axis) leading up to a band concert. What is the difference in the amount of time he spent on the final day versus the amount predicted by the line of best fit?

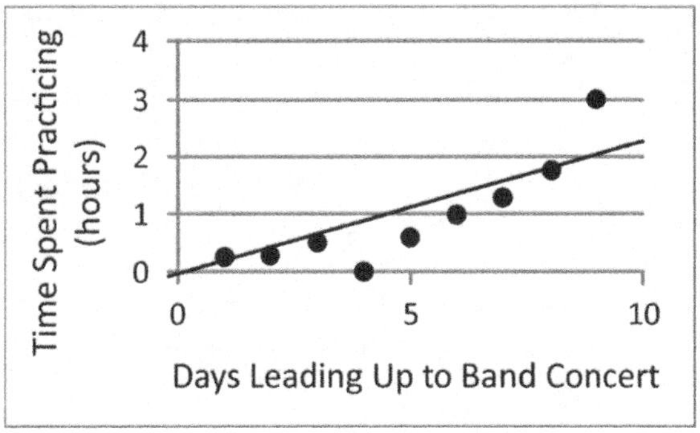

a. 0.5 hours
b. 1 hour
c. 2 hours
d. 3 hours

Copyright © Mometrix Media. You have been licensed one copy of this document for personal use only. Any other reproduction or redistribution is strictly prohibited. All rights reserved.
This content is provided for test preparation purposes only and does not imply an endorsement by Mometrix of any particular political, scientific, or religious point of view.

28. A function $f(x)$ is defined by $f(x) = 2x^2 + 7$. What is the value of $2f(x) - 3$?

a. $4x^2 + 11$
b. $4x^4 + 11$
c. $x^2 + 11$
d. $4x^2 + 14$

29. The equation of the line, in slope-intercept form, that passes through the points $(-2, -3)$ and $(4, -4)$ is __________.

a. $y = -\frac{1}{6}x + \frac{10}{3}$
b. $y = -\frac{1}{6}x - \frac{10}{3}$
c. $y = \frac{1}{6}x - \frac{10}{3}$
d. $y = -6x - \frac{10}{3}$

30. Determine which pair of equations are NOT inverses.

a. $y = x + 6; y = x - 6$
b. $y = 2x + 3; y = 2x - 3$
c. $y = \frac{2x+3}{x-1}; y = \frac{x+3}{x-2}$
d. $y = \frac{x-1}{2}; y = 2x + 1$

Copyright © Mometrix Media. You have been licensed one copy of this document for personal use only. Any other reproduction or redistribution is strictly prohibited. All rights reserved.
This content is provided for test preparation purposes only and does not imply an endorsement by Mometrix of any particular political, scientific, or religious point of view.

31. A school is selling tickets to its production of *Annie Get Your Gun*. Student tickets cost $3 each, and non-student tickets are $5 each. To offset the costs of the production, the school must earn at least $300 in ticket sales. Which graph shows the number of tickets the school must sell to offset production costs?

a.

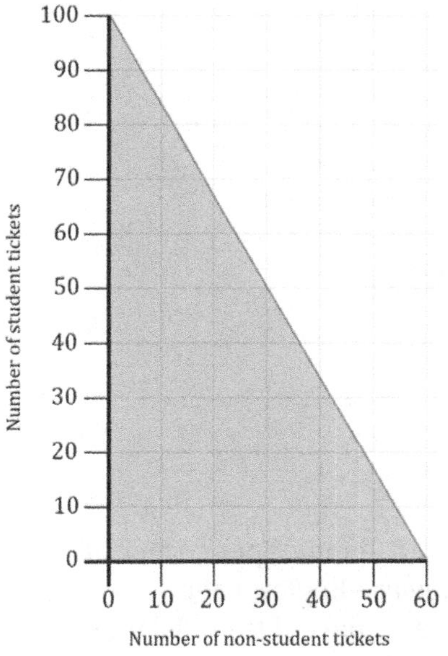

b.

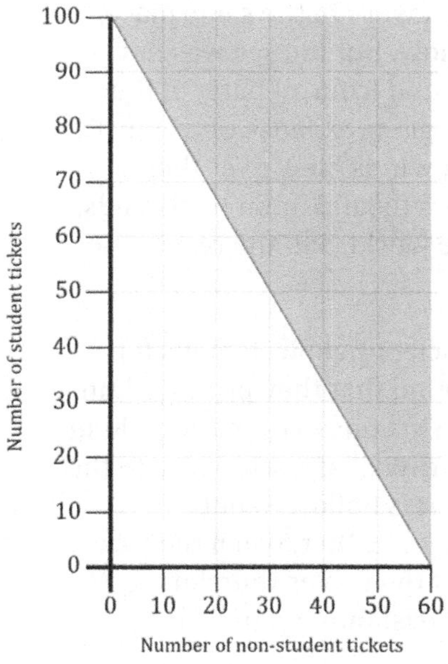

c.

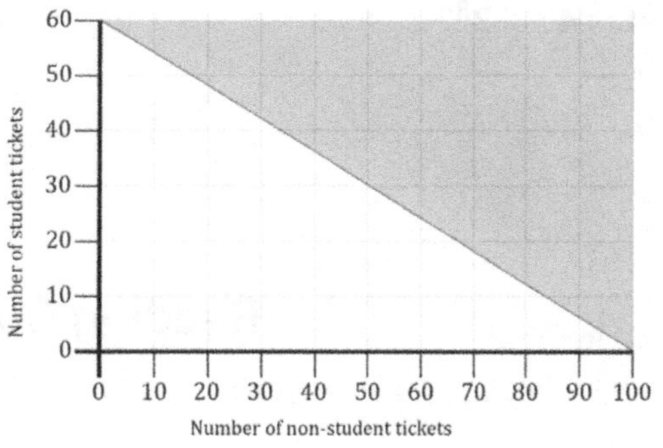

d.

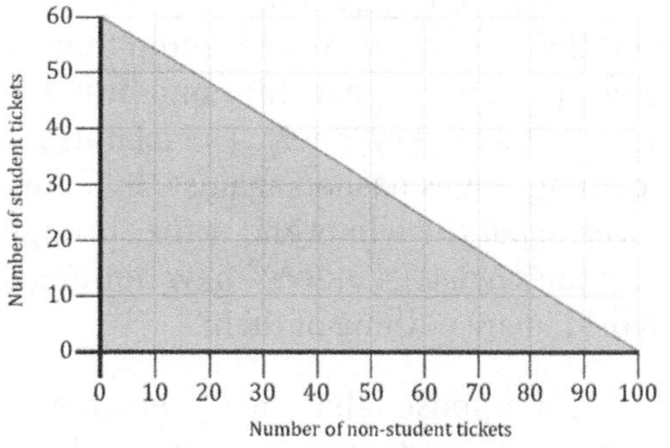

Copyright © Mometrix Media. You have been licensed one copy of this document for personal use only. Any other reproduction or redistribution is strictly prohibited. All rights reserved.
This content is provided for test preparation purposes only and does not imply an endorsement by Mometrix of any particular political, scientific, or religious point of view.

32. Complete the analogy: $\frac{1}{x}$ is to $\frac{1}{y}$ as...

a. x^2 is to y^3
b. e^x is to $\log y, y > 0$
c. x is to $\sqrt{y}, x, y \neq 0$
d. x^2 is to $\sqrt{y}, y > 0$

33. What is $4^6 \div 2^8$?

a. 2
b. 8
c. 16
d. 32

Reading Comprehension

Refer to the following for questions 1–6:

History of England

by Charles Dickens

If you look at a Map of the World, you will see, in the left-hand upper corner of the Eastern Hemisphere, two Islands lying in the sea. They are England and Scotland, and Ireland. England and Scotland form the greater part of these Islands. Ireland is the next in size. The little neighbouring islands, which are so small upon the Map as to be mere dots, are chiefly little bits of Scotland, —broken off, I dare say, in the course of a great length of time, by the power of the restless water.

In the old days, a long, long while ago..., these Islands were in the same place, and the stormy sea roared round them, just as it roars now. But the sea was not alive, then, with great ships and brave sailors, sailing to and from all parts of the world. It was very lonely. The Islands lay solitary, in the great expanse of water. The foaming waves dashed against their cliffs, and the bleak winds blew over their forests; but the winds and waves brought no adventurers to land upon the Islands, and the savage Islanders knew nothing of the rest of the world, and the rest of the world knew nothing of them.

It is supposed that the Phoenicians, who were an ancient people, famous for carrying on trade, came in ships to these Islands, and found that they produced tin and lead; both very useful things, as you know, and both produced to this very hour upon the sea-coast. The most celebrated tin mines in Cornwall are, still, close to the sea. One of them, which I have seen, is so close to it that it is hollowed out underneath the ocean; and the miners say, that in stormy weather, when they are at work down in that deep place, they can hear the noise of the waves thundering above their heads. So, the Phœnicians, coasting about the Islands, would come, without much difficulty, to where the tin and lead were.

The Phœnicians traded with the Islanders for these metals, and gave the Islanders some other useful things in exchange. The Islanders were, at first, poor savages, going almost naked, or only dressed in the rough skins of beasts, and staining their bodies, as other savages do, with coloured earths and the juices of

Copyright © Mometrix Media. You have been licensed one copy of this document for personal use only. Any other reproduction or redistribution is strictly prohibited. All rights reserved.
This content is provided for test preparation purposes only and does not imply an endorsement by Mometrix of any particular political, scientific, or religious point of view.

plants. But the Phœnicians, sailing over to the opposite coasts of France and Belgium, and saying to the people there, 'We have been to those white cliffs across the water, which you can see in fine weather, and from that country, which is called Britain, we bring this tin and lead,' tempted some of the French and Belgians to come over also. These people settled themselves on the south coast of England, which is now called Kent; and, although they were a rough people too, they taught the savage Britons some useful arts, and improved that part of the Islands. It is probable that other people came over from Spain to Ireland, and settled there.

Thus, by little and little, strangers became mixed with the Islanders, and the savage Britons grew into a wild, bold people; almost savage, still, especially in the interior of the country away from the sea where the foreign settlers seldom went; but hardy, brave, and strong.

The whole country was covered with forests, and swamps. The greater part of it was very misty and cold. There were no roads, no bridges, no streets, no houses that you would think deserving of the name. A town was nothing but a collection of straw-covered huts, hidden in a thick wood, with a ditch all round, and a low wall, made of mud, or the trunks of trees placed one upon another. The people planted little or no corn, but lived upon the flesh of their flocks and cattle. They made no coins, but used metal rings for money. They were clever in basket-work, as savage people often are; and they could make a coarse kind of cloth, and some very bad earthenware. But in building fortresses they were much more clever.

They made boats of basket-work, covered with the skins of animals, but seldom, if ever, ventured far from the shore. They made swords, of copper mixed with tin; but, these swords were of an awkward shape, and so soft that a heavy blow would bend one. They made light shields, short pointed daggers, and spears—which they jerked back after they had thrown them at an enemy, by a long strip of leather fastened to the stem. The butt-end was a rattle, to frighten an enemy's horse. The ancient Britons, being divided into as many as thirty or forty tribes, each commanded by its own little king, were constantly fighting with one another, as savage people usually do; and they always fought with these weapons.

1. According to the author, why did the ancient Britons regularly fight with each other?

a. They had many weapons.
b. They disliked the Phoenicians.
c. There were no roads or bridges.
d. They were divided into many tribes.

2. Which phrase best shows the change Dickens believes the Phœnicians caused in the ancient Britons?

a. "The Phoenicians traded with the Islanders..."
b. "They were clever in basket-work..."
c. "These people settled themselves on the south coast of England..."
d. "...the savage Britons grew into a wild, bold people..."

Copyright © Mometrix Media. You have been licensed one copy of this document for personal use only. Any other reproduction or redistribution is strictly prohibited. All rights reserved.
This content is provided for test preparation purposes only and does not imply an endorsement by Mometrix of any particular political, scientific, or religious point of view.

3. Which sentence or phrase best expresses the isolation of the islands of England and Scotland and Ireland?

a. "...which are so small upon the Map as to be mere dots"
b. "The Islands lay solitary, in the great expanse of water."
c. "they can hear the noise of the waves thundering above their heads."
d. "Thus, by little and little, strangers became mixed with the Islanders..."

4. Read this phrase from paragraph 7:

But seldom, if ever, ventured far from the shore.

Why does the author include this phrase?

a. To show the ways in which the people used boats
b. To highlight the irony of making boats
c. To show where the people used weapons
d. To explain why the people built fortresses

5. Why does the author begin the passage by describing a map?

a. To explain the location of the islands
b. To show the roads that run through England, Scotland, and Ireland
c. To show how the little bits of Scotland broke away from the main island
d. To show the size of the islands in relation to France and Belgium

6. What sentence or phrase best describes the lands in the interior of the islands? In other words, these lands are the parts away from the coast.

a. "These people settled themselves on the south coast of England, which is now called Kent..."
b. "Especially in the interior of the country away from the sea where the foreign settlers seldom went..."
c. The whole country was covered with forests and swamps.
d. "The ancient Britons, being divided into as many as thirty or forty tribes, each commanded by its own little king..."

Refer to the following for questions 7–12:

"The Gift of the Magi"

by O. Henry

[Jim and Della are a young husband and wife. They are very poor, and it is Christmas Eve. Della has been able to save $1.87 for a Christmas present. To buy a better gift for her husband, she decides to sell her beautiful hair to buy a fob for Jim. The fob will be for Jim's watch which is his most prized possession.]

When Della reached home her intoxication gave way a little to prudence and reason. She got out her curling irons and lighted the gas and went to work repairing the ravages made by generosity added to love. Which is always a tremendous task, dear friends--a mammoth task.

Within forty minutes her head was covered with tiny, close-lying curls that made her look wonderfully like a truant schoolboy. She looked at her reflection in the mirror long, carefully, and critically.

Copyright © Mometrix Media. You have been licensed one copy of this document for personal use only. Any other reproduction or redistribution is strictly prohibited. All rights reserved.
This content is provided for test preparation purposes only and does not imply an endorsement by Mometrix of any particular political, scientific, or religious point of view.

"If Jim doesn't kill me," she said to herself, "before he takes a second look at me, he'll say I look like a Coney Island chorus girl. But what could I do—oh! What could I do with a dollar and eighty-seven cents?"

At seven o'clock the coffee was made and the frying pan was on the back of the stove hot and ready to cook the chops.

Jim was never late. Della doubled the fob chain in her hand and sat on the corner of the table near the door that he always entered. Then she heard his step on the stairway down on the first flight, and she turned white for just a moment. She had a habit of saying little silent prayers about the simplest everyday things and now she whispered: "Please God, make him think I am still pretty."

* * *

Jim stopped inside the door, as immovable as a setter at the scent of quail. His eyes were fixed upon Della, and there was an expression in them that she could not read, and it terrified her. It was not anger, nor surprise, nor disapproval, nor horror, nor any of the sentiments that she had been prepared for. He simply stared at her fixedly with that peculiar expression on his face.

Della wriggled off the table and went for him.

"Jim, darling," she cried, "don't look at me that way. I had my hair cut off and sold it because I couldn't have lived through Christmas without giving you a present. It'll grow out again—you won't mind, will you? I just had to do it. My hair grows awfully fast. Say 'Merry Christmas!' Jim, and let's be happy. You don't know what a nice—what a beautiful, nice gift I've got for you."

"You've cut off your hair?" asked Jim, laboriously, as if he had not arrived at that patent fact yet even after the hardest mental labor.

"Cut it off and sold it," said Della. "Don't you like me just as well, anyhow? I'm me without my hair, ain't I?"

Jim looked about the room curiously.

"You say your hair is gone?" he said, with an air almost of idiocy.

"You needn't look for it," said Della. "It's sold, I tell you—sold and gone, too. It's Christmas Eve, boy. Be good to me, for it went for you. Maybe the hairs of my head were numbered," she went on with a sudden serious sweetness, "but nobody could ever count my love for you. Shall I put the chops on, Jim?"

Out of his trance Jim seemed quickly to wake. He enfolded his Della.

* * *

Jim drew a package from his overcoat pocket and threw it upon the table.

"Don't make any mistake, Dell," he said, "about me. I don't think there's anything in the way of a haircut or a shave or a shampoo that could make me like my girl any less. But if you'll unwrap that package you may see why you had me going a while at

Copyright © Mometrix Media. You have been licensed one copy of this document for personal use only. Any other reproduction or redistribution is strictly prohibited. All rights reserved.
This content is provided for test preparation purposes only and does not imply an endorsement by Mometrix of any particular political, scientific, or religious point of view.

first." White fingers and nimble tore at the string and paper. And then an ecstatic scream of joy; and then, alas! a quick, feminine change to hysterical tears and wails, necessitating the immediate employment of all the comforting powers of the lord of the flat.

For there lay the Combs—the set of combs, side and back, that Della had worshipped for long in a Broadway window. Beautiful combs, pure tortoise shell, with jeweled rims—just the shade to wear in the beautiful vanished hair. They were expensive combs, she knew, and her heart had simply craved and yearned over them without the least hope of possession. And now, they were hers, but the tresses that should have adorned the coveted adornments were gone. But she hugged them to her bosom, and at length she was able to look up with dim eyes and a smile and say: "My hair grows so fast, Jim!"

And then Della leaped up like a little singed cat and cried, "Oh, oh!"

Jim had not yet seen his beautiful present. She held it out to him eagerly upon her open palm. The dull precious metal seemed to flash with a reflection of her bright and ardent spirit.

"Isn't it a dandy, Jim? I hunted all over town to find it. You'll have to look at the time a hundred times a day now. Give me your watch. I want to see how it looks on it."

Instead of obeying, Jim tumbled down on the couch and put his hands under the back of his head and smiled.

"Della," said he, "let's put our Christmas presents away and keep 'em a while. They're too nice to use just at present. I sold the watch to get the money to buy your combs. And now suppose you put the chops on."

7. What is the meaning of the word *ardent*?

a. procrastinate
b. passionate
c. lukewarm
d. uncaring

8. What is the chronological order of these events: ?

(1) Jim gives Della's present to her | (2) Della prepares dinner | (3) Jim assures Della that he is not upset | (4) Della prepares her hair | (5) Della prays for Jim's reaction | (6) Jim receives his present from Della

a. 4, 3, 2, 1, 5, 6
b. 6, 1, 4, 3, 2, 5
c. 4, 2, 1, 5, 3, 6
d. 4, 2, 5, 3, 1, 6

Copyright © Mometrix Media. You have been licensed one copy of this document for personal use only. Any other reproduction or redistribution is strictly prohibited. All rights reserved.
This content is provided for test preparation purposes only and does not imply an endorsement by Mometrix of any particular political, scientific, or religious point of view.

9. In the passage, O. Henry uses several similes. Which of the following is not an example of a simile?

a. "her head was covered with tiny, close-lying curls that made her look wonderfully like a truant schoolboy."
b. "Jim stopped inside the door, as immovable as a setter at the scent of quail."
c. "'I hunted all over town to find it.'"
d. "And then Della leaped up like a little singed cat and cried, 'Oh, oh!'."

10. Which of the following best shows how O. Henry builds tension in the story?

a. "It was even worthy of The Watch. As soon as she saw it she knew that it must be Jim's."
b. "The door opened and Jim stepped in and closed it. He looked very thin and very serious."
c. "Jim looked about the room curiously. 'You say your hair is gone?' he said, with an air almost of idiocy."
d. "Jim was never late. Della doubled the fob chain in her hand and sat on the corner of the table near the door that he always entered."

11. What is the theme of this passage?

a. True love leads to the sacrifice of one's most precious possessions.
b. Christmas is the representation of the love and sacrifice needed by all.
c. The love between two people eventually leads to frustration.
d. People should not try to surprise each other gifts.

12. Who is the narrator of this passage?

a. Della
b. Jim
c. O. Henry
d. None of the above

Refer to the following for questions 13–18:

[In 1906, Elinore Pruitt Stewart moved to Denver for housework to support her daughter, Jerrine. Her employer in Denver was Mrs. Juliet Coney. A few years later, she moved to Wyoming to be a housekeeper for a rancher. The following passage is one of many letters that Stewart wrote to Mrs. Coney on life as a homesteader in Wyoming.]

A Letter of Elinore Pruitt Stewart

January 23, 1913

When I read of the hard times among the Denver poor, I feel like urging them every one to get out and file on land. I am very enthusiastic about women homesteading. It really requires less strength and labor to raise plenty to satisfy a large family than it does to go out to wash, with the added satisfaction of knowing that their job will not be lost to them if they care to keep it. Even if improving the place does go slowly, it is that much done to stay done. Whatever is raised is the homesteader's own, and there is no house-rent to pay. This year Jerrine cut and dropped enough potatoes to raise a ton of fine potatoes. She wanted to try, so we let her, and you will remember that she is but six years old.... Any woman strong enough to go out by the day could have done every bit of the work and put in two or three times that much, and it would have been so much more pleasant than to work so hard in the city and be on starvation rations all winter.

Copyright © Mometrix Media. You have been licensed one copy of this document for personal use only. Any other reproduction or redistribution is strictly prohibited. All rights reserved.
This content is provided for test preparation purposes only and does not imply an endorsement by Mometrix of any particular political, scientific, or religious point of view.

To me, homesteading is the solution of all poverty's problems, but I realize that temperament has much to do with success in any undertaking, and persons afraid of coyotes and work and loneliness had better let ranching alone. At the same time, any woman who can stand her own company, can see the beauty of the sunset, loves growing things, and is willing to put in as much time at careful labor as she does over the washtub, will certainly succeed; will have independence, plenty to eat all the time, and a home of her own in the end.

Experimenting need cost the homesteader no more than the work, because by applying to the Department of Agriculture at Washington he can get enough of any seed and as many kinds as he wants to make a thorough trial, and it doesn't even cost postage. Also one can always get bulletins from there and from the Experiment Station of one's own State concerning any problem or as many problems as may come up. I would not, for anything, allow Mr. Stewart to do anything toward improving my place, for I want the fun and the experience myself. And I want to be able to speak from experience when I tell others what they can do. Theories are very beautiful, but facts are what must be had, and what I intend to give some time.

13. The writer of this letter is suggesting that women should own land and farm rather than

a. cook in a restaurant.
b. open a bed and breakfast.
c. do laundry for others.
d. teach in a one-room schoolhouse.

14. Stewart mentions her daughter's potato crop. She does this to show

a. that child labor is acceptable.
b. that there are no schools in the area.
c. that women work just as hard as men do.
d. how easy it is to raise crops.

15. What do you think Mrs. Coney's reaction to the letter might have been?

a. She was probably glad to be rid of such a lazy worker.
b. She may be glad to know that Mrs. Stewart is enjoying her time with homesteading.
c. She may have been sorry that she too did not homestead.
d. She was likely angry that Mrs. Stewart had written.

16. Which of the following does Stewart NOT give as an advantage of homesteading?

a. It takes less strength and work than doing laundry for others
b. The worker cannot lose her job if she wants to keep it.
c. No one has to pay rent.
d. One can always find good company.

17. Which of the following is a risk for the poor in Denver?

a. The possibility of losing their jobs
b. The likelihood of a strike
c. The probability of a landslide
d. Losing their homes to fire

Copyright © Mometrix Media. You have been licensed one copy of this document for personal use only. Any other reproduction or redistribution is strictly prohibited. All rights reserved.
This content is provided for test preparation purposes only and does not imply an endorsement by Mometrix of any particular political, scientific, or religious point of view.

18. The tone of the letter is

a. complaining and bitter.
b. sad and lonely.
c. positive and encouraging.
d. hopeless and despairing.

Refer to the following for questions 19–24:

Forest Manager

Salvage logging is removing dead or dying forest stands that are left behind by a fire or disease. This practice has been used for several decades. These dead or dying trees become fuel that feeds future fires. The best way to lower the risk of forest fires is to remove the dead timber from the forest floor. Salvage logging followed by replanting ensures the reestablishment of desirable tree species.

For example, planting conifers accelerates the return of fire-resistant forests. Harvesting timber helps forests by reducing fuel load, thinning the forest stands, and relieving competition between trees. Burned landscapes leave black surfaces and ash layers that have very high soil temperatures. These high soil temperatures can kill many plant species. Logging mixes the soil. So, this lowers surface temperatures to more normal levels. The shade from material that is left behind by logging also helps to lower surface temperatures. After an area has been salvage logged, seedlings in the area start to grow almost immediately. However, this regrowth can take several years in areas that are not managed well.

Ecology Professor

Salvage logging moves material like small, broken branches to the forest floor. These pieces can become fuel for more fires. The removal of larger, less flammable trees leaves behind small limbs and increases the risk of forest fires. In unmanaged areas, these pieces are found more commonly on the tops of trees where they are unavailable to fires. Logging destroys old forests that are more resistant to wildfires. So, this creates younger forests that are more open to fires. In old forests, branches of bigger trees are higher above the floor where fires may not reach.

Replanting after wildfires creates monoculture plantations where only a single crop is planted. This monoculture allows less biological diversity. Also, it allows plants to be less resistant to disease. So, this increases the chance of fire. Salvage logging also upsets natural forest regrowth by killing most of the seedlings that grow after a wildfire. It breaks up the soil and increases erosion. Also, it removes most of the shade that is needed for young seedlings to grow.

19. According to the professor, why are the broken branches in unmanaged forests preferable to those in logged areas for wildfire resistance?

a. They are left on the forest floor and bring nutrients to the soil.
b. They are left on the forest floor and serve as fuel for fires.
c. They are left on the tops of trees where fires cannot reach.
d. They are spread more evenly across the forest floor.

Copyright © Mometrix Media. You have been licensed one copy of this document for personal use only. Any other reproduction or redistribution is strictly prohibited. All rights reserved.
This content is provided for test preparation purposes only and does not imply an endorsement by Mometrix of any particular political, scientific, or religious point of view.

20. Which of the following is NOT a supporting detail for the forest manager's argument?

a. "This practice has been used for decades."
b. "Logging mixes the soil. So, this lowers surface temperatures to more normal levels."
c. "After an area has been salvage logged, seedlings in the area start to grow almost immediately."
d. "Salvage logging is removing dead or dying forest stands that are left behind by a fire or disease."

21. A study compared two plots of land that were managed differently after a fire. Plot A was salvage logged. Plot B was left unmanaged. After a second fire, they compared two plant groups between Plots A and B. They found that both plant groups burned worse in Plot A than in Plot B. Whose viewpoint do these results support?

a. Only the manager
b. Only the professor
c. Both the manager and professor
d. Neither the manager nor the professor

22. What is the main idea of the forest manager's argument?

a. Salvage logging is helpful because it removes dead or dying timber from the forest floor. So, this lowers the risk of future fires.
b. Salvage logging is helpful because it has been practiced for many decades.
c. Salvage logging is harmful because it raises soil temperatures above normal levels. So, this threatens the health of plant species.
d. Salvage logging is helpful because it gives shade for seedlings to grow after a wildfire.

23. Which of the following statements does NOT agree with the professor?

a. In younger forests, small branches are closer to the forest floor and more available for fires.
b. Old growth forests have taller trees, so branches are high up and fires cannot reach them.
c. Monoculture forests have less biological diversity and fewer disease-resistant trees.
d. Larger trees are common in old growth forests and serve as the main fuel source for fires.

24. Whose viewpoints would potentially be confirmed by a future study looking at the spreading out and regrowth of seedlings for many years after a wildfire in managed and unmanaged forests?

a. Only the manager
b. Only the professor
c. Both the manager and professor
d. Neither the manager nor professor

Refer to the following for questions 25–29:

Section 1: Improving Diets

A healthier diet is something that many people want for themselves. However, this can be a struggle to put into practice for many people. This does not mean that just because it's hard and frustrating doesn't mean that people should stop trying.

A powerful and easy approach to improving diets is to know that some foods are so good for us that we can almost think of them as medicine. Some foods help to fight heart disease, cancer, or depression. Other foods help to lower cholesterol or

Copyright © Mometrix Media. You have been licensed one copy of this document for personal use only. Any other reproduction or redistribution is strictly prohibited. All rights reserved.
This content is provided for test preparation purposes only and does not imply an endorsement by Mometrix of any particular political, scientific, or religious point of view.

blood pressure. Broccoli is high in vitamin K and vitamin C which help build strong bones and fight off cancers. Avocadoes can lower cholesterol and help reduce the risk of heart disease. Sweet potatoes are full of cancer-fighting and immune system-boosting vitamin A. Garlic can slow down the growth of bacteria and has been shown to lower cholesterol and blood pressure. Spinach is a great cancer fighter and has immune-boosting antioxidants important for eye health. Beans help lower risk of heart disease and breast cancer.

At some point, people want to give themselves the full treatment: diet, exercise, and general health overhaul. In the meantime, they can take the baby step of adding in one or more healthy food a week. This step is quick, easy, and painless. It couldn't be simpler to implement. Also, it will make their switch to healthy eating much easier to accomplish when they finally get there.

Section 2: Dietary Guidelines for Americans

The Dietary Guidelines for Americans is put together by the US Department of Health and Human Services and the US Department of Agriculture. The guidelines offer advice to people about food choices that advance good health and lower the risk of certain diseases (e.g., hypertension, anemia, and osteoporosis). In addition, this form offers a detailed outline on the kinds of foods that people should have in their diets. The outline is given so that additional supplements or vitamins may not be necessary. The form also has information on the types of exercise that are necessary for someone to stay healthy. Also, there is information on to handle and prepare certain foods to lower the risk of foodborne illness.

The Food Pyramid gave recommendations for the number of daily servings from each group. The USDA's Food Pyramid was heavily criticized for being unclear and confusing. In 2011, MyPlate replaced the Food Pyramid. MyPlate is much easier to understand because it has a picture of a dinner plate that is divided into four sections. So, this shows how our daily diet should be spread out among the different food groups. Vegetables and grains each take up 30% of the plate. Fruits and proteins each make up 20% of the plate. In the corner of the image is a cup that is marked as Dairy.

Most experts consider MyPlate to be a great improvement over the Food Pyramid. However, some it has still come under criticism from some quarters. Many believe too much emphasis is placed on protein, and some say the dairy recommendation should be eliminated altogether. The Harvard School of Public Health created its own Healthy Eating Plate to address what it sees as shortcomings in MyPlate. Harvard's guide adds healthy plant-based oils to the mix, stresses whole grains instead of merely grains, recommends drinking water or unsweetened coffee or tea instead of milk, and adds a reminder that physical activity is important.

Section 3: Preparing Better Meals in the Food Industry

People in the food industry that want to prepare a healthy meal for their customers should first decide on the nutritional goals of their menu. Once these goals have been set up, you should continue to plan by researching foods. These foods need to meet your goals without going beyond the available time and resource limits. Then, you can put together a meal plan that list several details. These details should have what foods will be included, the

Copyright © Mometrix Media. You have been licensed one copy of this document for personal use only. Any other reproduction or redistribution is strictly prohibited. All rights reserved.
This content is provided for test preparation purposes only and does not imply an endorsement by Mometrix of any particular political, scientific, or religious point of view.

average time it takes to prepare and cook each of these meals, and the cost of preparing these meals. The next step is to decide on the best way of preparing the food for these meals. Think about which foods should be prepared first and the best ways to handle or prepare your food to lower the risk of illness. Also, think about methods that can be used to lower the cooking time. Finally, you can prepare the meal according to your plans.

When you need to decide on what foods to prepare, you need to think about several things. You should consider the food's nutritional value, the time it takes to prepare each food, the number of people to be served, and the cost of preparing each food. Each food has its own cooking time and has different nutrients. So, it is important to prepare foods that meet people's nutritional goals without using too much time for cooking the meal. Since you will likely have a budget for the meal, you need to review the number of people to be served and the cost of preparing each food. If the cost is too high, some meals may not be good choices to serve to large groups. For example, you are interested in serving a good source of protein for a meal. So, steak may be a good option for a small group of people. However, that would probably be too expensive for a larger group.

25. What is the main idea of Section 1?

a. Preparing a menu requires thorough research.
b. Some foods are healthier than others.
c. Positive diet changes can be simple.
d. Some people can make dietary changes and some cannot.

26. What is the purpose of including Section 2 in this passage?

a. To explain how the Food Pyramid was poorly designed
b. To cover the government's influence on the dietary recommendations for Americans
c. To show that there is no perfect system for coming up with dietary recommendations
d. To share information on generally accepted nutrition guidelines

27. What is the purpose of including Harvard's Healthy Eating Plate in Section 2?

a. To support the assertion MyPlate has received criticism
b. To suggest Healthy Eating Plate may replace current USDA dietary guidelines
c. It shows another option as a dietary guideline
d. To highlight the influence of an Ivy League school

28. According to the text, when preparing food for customers, all of the following are important EXCEPT which?

a. Foodborne illness
b. Calories in a meal
c. Layering the flavors in a dish
d. Preparation time

29. What is the tone of the three sections?

a. Condemning
b. Informative
c. Serious
d. Pretentious

Copyright © Mometrix Media. You have been licensed one copy of this document for personal use only. Any other reproduction or redistribution is strictly prohibited. All rights reserved.
This content is provided for test preparation purposes only and does not imply an endorsement by Mometrix of any particular political, scientific, or religious point of view.

Refer to the following for questions 30–36:

Excerpt of a judicial review summary written by Bernard Schwartz

Judicial review, the power of courts to determine the legality of governmental acts, usually refers to the authority of judges to decide a law's constitutionality. Although state courts exercised judicial review prior to the ratification of the Constitution, the doctrine is most often traced to the landmark US Supreme Court decision Marbury v. Madison (1803), which struck down an act of Congress as unconstitutional. In a now classic opinion, Chief Justice John Marshall found the power of judicial review implied in the Constitution's status as "the supreme Law of the Land" prevailing over ordinary laws.

Both federal and state courts have exercised judicial review. Federal courts review federal and state acts to ensure their conformity to the Constitution and the supremacy of federal over state law; state courts review laws to ensure their conformity to the US Constitution and their own state constitutions. The power of judicial review can be exercised by any court in which a constitutional issue arises.

Judicial review gained added importance in the late nineteenth and early twentieth centuries, as courts passed judgement on laws regulating corporate behavior and working conditions. In these years, the Supreme Court repeatedly struck down laws regulating wages, hours of labor, and safety standards. This is often called the Lochner Era, after Lochner v. New York, a 1905 decision ruling a New York maximum-hours law unconstitutional on the grounds that it violated the Fourteenth Amendment. During this period, the Supreme Court invalidated no fewer than 228 state laws.

Justice Oliver Wendell Holmes Jr., dissenting from many of these decisions, urged judges to defer to legislatures. In the later 1930's, the Supreme Court adopted the Holmes approach-partly in response to the threat of President Franklin Delano Roosevelt's "court packing" plan of 1937. Deferring to legislative judgement, the Supreme Court thereafter upheld virtually all laws regulating business and property rights, including laws similar to those invalidated during the Lochner Era.

Under the chief justiceship of Earl Warren (1953-1969) and beyond, however, the Court moved toward striking down law restricting personal rights and liberties guaranteed by the Bill of Rights, particularly measures limiting freedom of expression, freedom or religion, the right of criminal defendants, equal treatment of the sexes, and the rights of minorities to equal protection of the law. In another extension of judicial review, the Court read new rights into the Constitution, notably the right of privacy (including abortion rights) and invalidated laws restricting those rights. Many other countries including Germany, Italy, France, and Japan, adopted the principle of judicial review after World War II, making constitutional law one the more important recent American exports.

30. Which of the following statements about judicial review does the passage best support?

a. States should defer to the Federal Government when interpreting the Constitution.
b. Judicial Review was started due to the Lochner Era.
c. The Constitution overrides state law in some cases.
d. Judicial Review was founded by Earl Warren Chief Justice of the Supreme Court.

Copyright © Mometrix Media. You have been licensed one copy of this document for personal use only. Any other reproduction or redistribution is strictly prohibited. All rights reserved.
This content is provided for test preparation purposes only and does not imply an endorsement by Mometrix of any particular political, scientific, or religious point of view.

31. From the passage, it can be inferred ordinary laws created by lawmakers must be within the framework of the Constitution. Which of the following sentences supports this claim the best?

a. Although state courts exercised...
b. Both federal and state courts...
c. Judicial review gained added importance...
d. During this period, the Supreme...

32. Which of the following words best characterizes the content of the passage?

a. Historical
b. Introspective
c. Prospective
d. Demonstrative

33. The word *ratification* as used in this passage refers to

a. endorsement
b. disapprove
c. limiting
d. embargo

34. The word *dissenting* as used in this passage refers to

a. headstrong
b. compatible
c. obliging
d. contradictory

35. The explanation of judicial review is based on

a. Constitutional arguments
b. Chief Justice Warren
c. Marbury v Madison
d. Lochner v New York

36. What was Franklin Delano Roosevelt's contribution to the process of Judicial Review according to the summary?

a. The appointment of Earl Warren to the Court
b. Adopted the Holmes approach according to the author
c. Threatened the Court according to the author
d. Upheld laws regulating business

Copyright © Mometrix Media. You have been licensed one copy of this document for personal use only. Any other reproduction or redistribution is strictly prohibited. All rights reserved.
This content is provided for test preparation purposes only and does not imply an endorsement by Mometrix of any particular political, scientific, or religious point of view.

Answer Key and Explanations #1

Algebra and Functions

1. C: A system of linear equations can be solved by using matrices or by using the graphing, substitution, or elimination (also called linear combination) method. The elimination method is shown here:

$$\begin{aligned} 3x + 4y &= 2 \\ 2x + 6y &= -2 \end{aligned}$$

In order to eliminate x by linear combination, multiply the top equation by 2 and the bottom equation by –3 so that the coefficients of the x-terms will be additive inverses.

$$\begin{aligned} 2(3x + 4y &= 2) \\ -3(2x + 6y &= -2) \end{aligned}$$

Then, add the two equations and solve for y.

$$\begin{aligned} 6x + 8y &= 4 \\ \underline{-6x - 18y} &\underline{= 6} \\ -10y &= 10 \\ y &= -1 \end{aligned}$$

Substitute –1 for y in either of the given equations and solve for x.

$$\begin{aligned} 3x + 4y &= 2 \\ 3x + 4(-1) &= 2 \\ 3x - 4 &= 2 \\ 3x &= 6 \\ x &= 2 \end{aligned}$$

The solution to the system of equations is $(2, -1)$.

2. B: The inequality specifies that the difference between L and 15 inches must be less or equal to 0.01. Note that $|14.99 - 15| = |-0.01| = 0.01$, which is equal to the specified tolerance and therefore meets the condition.

3. D: First, state the exclusions of the domain.

$$\begin{gathered} x^3 - 7x + 6 \neq 0 \\ (x+3)(x-1)(x-2) \neq 0 \\ x + 3 \neq 0 \quad x - 1 \neq 0 \quad x - 2 \neq 0 \\ x \neq -3 \qquad x \neq 1 \qquad x \neq 2 \end{gathered}$$

To determine whether there are asymptotes or holes at these values of x, simplify the expression:

$$\frac{x^2 - x - 2}{x^3 - 7x + 6} = \frac{(x-2)(x+1)}{(x+3)(x-1)(x-2)} = \frac{x+1}{(x+3)(x-1)}$$

There are asymptotes at $x = 1$ and at $x = -3$ and a hole at $x = 2$. Statement I is false.

Copyright © Mometrix Media. You have been licensed one copy of this document for personal use only. Any other reproduction or redistribution is strictly prohibited. All rights reserved.
This content is provided for test preparation purposes only and does not imply an endorsement by Mometrix of any particular political, scientific, or religious point of view.

To find the x-intercept of $f(x)$, solve $f(x) = 0$. $f(x) = 0$ when the numerator is equal to zero. The numerator equals zero when $x = 2$ and $x = -1$; however, 2 is excluded from the domain of $f(x)$, so the x-intercept is –1. This is enough to say statement II is false. To find the y-intercept of $f(x)$, find $f(0)$.

$$f(0) = \frac{0^2 - 0 - 2}{0^3 - 7(0) + 6} = \frac{-2}{6} = -\frac{1}{3}$$

The y-intercept is $-\frac{1}{3}$, but the x-intercept is not $-\frac{1}{3}$. Statement II is false.

4. D: When asked to find the point of intersection of two lines, we are being asked to solve a system of equations. This system can be solved using the substitution method by substituting $x - 5$ for y in the first equation, $y = 2x + 3$.

$$x - 5 = 2x + 3$$

From here, solve for x.

$$-x - 5 = 3$$
$$-x = 8$$
$$x = -8$$

Now that the value of x is known, substitute this into either original equation and solve for y.

$$y = 2(-8) + 3 = -16 + 3 = -13$$

Therefore, the point of intersection for these two lines is $(-8, -13)$.

5. A: The value of the fraction $\frac{7}{5}$ can be evaluated by dividing 7 by 5, which yields 1.4. The average of 1.4 and 1.4 is $\frac{1.4+1.4}{2} = 1.4$.

6. C: According to the exponent rule $a^{-n} = \frac{1}{a^n}$, the expression 2^{-3} is equivalent to $\frac{1}{2^3}$. Since $2^3 = 2 \times 2 \times 2 = 8$, this expression is equivalent to $\frac{1}{8}$.

7. B: Graphically, a relationship between x and y can be identified as a function if the graph passes the vertical line test. Choice B is the graph of $y = x^3$, which is a function because it passes the vertical line test. A function is a relationship in which for every element of the domain (x) produces exactly one element of the range (y).

The first relation is a circle, which fails the vertical line test for functions. Choice C is a parabola on its side, which also fails the vertical line test and is therefore not a function. The relation in choice D has an element of the domain (2) that maps to two different range values, so it is also not a function.

8. D: Each hour has 60 minutes, and each of those minutes has 60 seconds.

$$60 \times 60 = 3{,}600$$

This means there are 3,600 seconds in 1 hour. To represent this fractional part of an hour, divide the part by the whole.

Copyright © Mometrix Media. You have been licensed one copy of this document for personal use only. Any other reproduction or redistribution is strictly prohibited. All rights reserved.
This content is provided for test preparation purposes only and does not imply an endorsement by Mometrix of any particular political, scientific, or religious point of view.

$$\frac{400}{3{,}600} = \frac{4}{36} = \frac{1}{9}$$

Therefore, 400 seconds is $\frac{1}{9}$ of an hour.

9. B: A test of each option at $x = -3$, $x = 0$, and $x = 4$ demonstrates that choice B is the closest representation.

x	**Observed**	y_a	y_b	y_c	y_d
-3	$\sim\frac{3}{4}$	$\frac{1}{3}$	$\frac{2}{3}$	$-\frac{7}{3}$	$\frac{1}{9}$
0	$\sim-\frac{3}{4}$	$-\frac{2}{3}$	$-\frac{2}{3}$	$\frac{1}{3}$	$-\frac{4}{3}$
4	$\sim\frac{3}{4}$	2	$\frac{2}{3}$	$\frac{7}{3}$	$\frac{8}{9}$

10. D: Consider each answer choice. The x-intercept is the point at which $f(x) = 0$. When $0 = \log_b x$, $b^0 = x$; since $b^0 = 1$, the x-intercept of $f(x) = \log_b x$ is always 1. If $f(x) = \log_b x$ and $x = b$, then $f(x) = \log_b b$, which is, by definition, 1 ($b^1 = b$). The statement $f(x) < 0$ is true only for x-values between 0 and 1. Choice D states, "If $g(x) = x^b$, the graph of $f(x)$ is symmetric to the graph of $g(x)$ with respect to $y = x$." This is a false statement. On the other hand, if $g(x) = b^x$, then $f(x)$ and $g(x)$ are inverse functions and are therefore symmetric with respect to $y = x$.

11. C: Since the rate, in miles per minute, is constant, this can be solved by setting up a proportion.

$$\frac{\text{miles}}{\text{min}} = \frac{10}{12} = \frac{210}{t}$$

Now, solve for t by using cross multiplication.

$$t = \frac{210 \times 12}{10} = 252$$

Mark will take 252 minutes to reach his destination. From here, convert the time to hours and minutes. Start by dividing 252 by 60.

$$252 \div 60 = 4.2$$

He will take 4 hours and 0.2 of another hour. To convert the partial hour, 0.2, to minutes, multiply it by 60.

$$0.2 \times 60 = 12$$

Therefore, it will take him 4 hours and 12 minutes to reach his destination.

12. D: When a number is raised to a power, you multiply the number by itself the number of times indicated by the power. For example, $2^3 = 2 \times 2 \times 2 = 8$. A number raised to the power of 0 is always equal to 1. So, 6^0 is the smallest number shown. Similarly, for the other numbers:

$$9 = 9;\ 10^1 = 10;\ 4^2 = 4 \times 4 = 16$$

Since $1 < 8 < 9 < 10 < 16$, we can write the order as $6^0, 2^3, 9, 10^1, 4^2$.

Copyright © Mometrix Media. You have been licensed one copy of this document for personal use only. Any other reproduction or redistribution is strictly prohibited. All rights reserved.
This content is provided for test preparation purposes only and does not imply an endorsement by Mometrix of any particular political, scientific, or religious point of view.

13. D: Manipulate the inequality to isolate the variable. Start by adding 12 to each side.

$$4x - 12 < 4$$
$$4x - 12 + 12 < 4 + 12$$
$$4x < 16$$

Then divide both sides by 4.

$$\frac{4x}{4} < \frac{16}{4}$$
$$x < 4$$

Since x must be less than and not equal to 4, only choice D works.

14. D: Rafael's profit on each computer is given by the difference between the price he pays and the price he charges his customer, or \$800 − \$450. If he sells n computers in a month, his total profit will be n times this difference, or $n(\$800 - \$450)$. However, it is necessary to subtract his fixed costs of \$3,000 from this to compute his final profit per month. This gives the complete equation:

$$P = n(\$800 - \$450) - \$3{,}000$$

15. D: Bacterial growth is exponential. Let x be the number of times the population doubles, a be the number of bacteria in the colony originally transferred into the broth, and y be the number of bacteria in the broth after x doubling times. After 1 hour, the population would have doubled three times:

$$a(2)^x = y$$
$$a(2^3) = 8 \times 10^6$$
$$8a = 8 \times 10^6$$
$$a = 10^6$$

So, the number of bacteria originally transferred into the petri dish was 10^6. The equation for determining the number of bacteria is $y = (2^x) \times 10^6$. Since the bacteria double every twenty minutes, they go through three doubling times every hour. So, when the bacteria are allowed to grow for eight hours, they will have gone through $8 \times 3 = 24$ doubling times. When $x = 24$, $y = (2^{24}) \times 10^6 = 16{,}777{,}216 \times 10^6$, which is approximately 1.7×10^{13}.

16. B: The vertical operators indicate absolute values, which are always positive. Start by simplifying the expressions inside the absolute value bars.

$$|7 - 5| - |5 - 7|$$
$$|2| - |-2|$$

Then, evaluate the absolute values and subtract. Since absolute value is always positive, both $|2|$ and $|-2|$ are equal to 2.

$$2 - 2 = 0$$

17. B: The path of a bullet is a parabola, which is the graph of a quadratic function. The path of a sound wave can be modeled by a sine or cosine function. The distance an object travels over time

Copyright © Mometrix Media. You have been licensed one copy of this document for personal use only. Any other reproduction or redistribution is strictly prohibited. All rights reserved.
This content is provided for test preparation purposes only and does not imply an endorsement by Mometrix of any particular political, scientific, or religious point of view.

given a constant rate is a linear relationship, while radioactive decay is modeled by an exponential function.

18. D: The graph shows five days that Jen runs. On three of the days (Tuesday, Thursday, and Friday), she runs four or more miles. So three out of five days, or $\frac{3}{5}$ of the time, she runs at least four miles.

19. B: First, set the equation equal to zero, then factor it.

$$4x^3y^2 - 16x^2y = -16x^3y + 4x^2y^2$$
$$4x^3y^2 + 16x^3y - 4x^2y^2 - 16x^2y = 0$$
$$4x^2y(xy + 4x - y - 4) = 0$$
$$4x^2y[x(y+4) - 1(y+4)] = 0$$
$$4x^2y[(y+4)(x-1)] = 0$$

Use the zero-product property to find the solution. Since the question only asks to solve for x, only look at the terms that have x in them.

$$4x^2y = 0 \qquad x - 1 = 0$$
$$x = 0 \qquad x = 1$$

So, the solutions are $x = \{0,1\}$.

20. D: Distance is the product of velocity and time.

$$\begin{aligned}(5 \times 10^6) \times (2 \times 10^{-4}) &= (5 \times 2) \times (10^6 \times 10^{-4}) \\ &= 10 \times 10^{6-4} \\ &= 10 \times 10^2 \\ &= 10^3 \\ &= 1{,}000\end{aligned}$$

Therefore, the bullet traveled 1,000 feet.

21. A: A simple test of each option at $x = -2$, $x = 0$, and $x = 1$ demonstrates that only choice A works:

x	**Observed**	y_a	y_b	y_c
−2	**~1**	$\frac{3}{4}$	$-\frac{3}{4}$	$-\frac{5}{4}$
0	$-\frac{\mathbf{3}}{\mathbf{2}}$	$-\frac{3}{2}$	$-\frac{3}{2}$	$\frac{1}{2}$
1	$-\frac{\mathbf{3}}{\mathbf{2}}$	$-\frac{3}{2}$	−3	$-\frac{1}{2}$

22. A: The x-intercept is the point at which $f(x) = 0$. When $0 = \log_b x$, $b^0 = x$; since $b^0 = 1$, the x-intercept of $f(x) = \log_b x$ is always 1. If $f(x) = \log_b x$ and $x = b$, then $f(x) = \log_b b$, which is, by definition, 1 ($b^1 = b$). The statement $f(x) < 0$ is true only for x-values between 0 and 1($0 < x < 1$). If $g(x) = b^x$, then $f(x)$ and $g(x)$ are inverse functions and are therefore symmetric with respect to $y = x$.

Copyright © Mometrix Media. You have been licensed one copy of this document for personal use only. Any other reproduction or redistribution is strictly prohibited. All rights reserved.
This content is provided for test preparation purposes only and does not imply an endorsement by Mometrix of any particular political, scientific, or religious point of view.

23. A: Marcus needs $325 for the new gaming system. If he earns $6.50 an hour, the number of hours he needs to work can be determined by dividing $325 by $6.50 which is written as $\frac{325}{6.50} = x$. Multiplying both sides of the equation by 6.50 yields $6.50x = 325$.

24. A: First, use the table of values to write polynomial equations for $f(x)$ and $g(x)$: $f(x) = 3x^2$ and $g(x) = 2x - 2$. Then, use those equations to find $f(g(-3))$.

$$g(-3) = 2(-3) - 2 = -8$$
$$f(-8) = 3(-8)^2 = 192$$

So, $f(g(-3)) = 192$.

25. A: Substitute the given values for a and b and then simplify using the order of operations.

$$\begin{aligned} a^2 + 3ab - b^2 &= (3)^2 + 3(3)(-2) - (-2)^2 \\ &= 9 + 3(3)(-2) - 4 \\ &= 9 - 18 - 4 \\ &= -13 \end{aligned}$$

Therefore, the value of the expression is −13.

26. A: The path of a sound wave follows the oscillations of a sinusoidal function (either a sine or cosine function can be used).

27. B: The point marked on the scatterplot for the final day shows that Julio practiced for 3 hours, while the line of best fit crosses the 2-hour mark on the same day. The difference between 3 hours and 2 hours is 1 hour.

28. A: Start by substituting the function $f(x)$ into the expression $2f(x) - 3$.

$$2(2x^2 + 7) - 3$$

From here, simplify the expression using the distributive property.

$$4x^2 + 14 - 3$$

Finally, combine like terms.

$$4x^2 + 11$$

29. B: The equation of the line can be expressed in slope-intercept form, $y = mx + b$, when m and b are determined. Start by solving for m, the slope, using the slope formula: $m = \frac{y_2 - y_1}{x_2 - x_1}$. Use the points $(-2, -3)$ and $(4, -4)$ to substitute x- and y-values into the formula.

$$m = \frac{y_2 - y_1}{x_2 - x_1} = \frac{-4 - (-3)}{4 - (-2)} = -\frac{1}{6}$$

Now solve for b, the y-intercept. Substitute in values for x, y, and m into the slope-intercept formula, $y = mx + b$. Use either pair of (x, y) coordinates. For example, when $(-2, -3)$ is used, $y = mx + b$ becomes $-3 = \left(-\frac{1}{6}\right)(-2) + b$. Now solve for the value of b.

Copyright © Mometrix Media. You have been licensed one copy of this document for personal use only. Any other reproduction or redistribution is strictly prohibited. All rights reserved.
This content is provided for test preparation purposes only and does not imply an endorsement by Mometrix of any particular political, scientific, or religious point of view.

$$-3 = \left(-\frac{1}{6}\right)(-2) + b$$
$$-3 = \frac{2}{6} + b$$
$$-\frac{10}{3} = b$$

We know that $m = -\frac{1}{6}$ and $b = -\frac{10}{3}$. Now the equation of the line can be expressed in slope-intercept form.

$$y = -\frac{1}{6}x - \frac{10}{3}$$

30. B: To find the inverse of an equation, solve for x in terms of y; then, exchange the variables x and y. Or, to determine if two functions $f(x)$ and $g(x)$ are inverses, find $f(g(x))$ and $g(f(x))$; if both results are x, then $f(x)$ and $g(x)$ are inverse functions.

For example, to find the inverse of $y = x + 6$, rewrite the equation $x = y + 6$ and solve for y. Since $y = x - 6$, the two given equations given in Choice A are inverses. Likewise, to find the inverse of $y = \frac{2x+3}{x-1}$, rewrite the equation as $x = \frac{2y+3}{y-1}$ and solve for y:

$$xy - x = 2y + 3$$
$$xy - 2y = x + 3$$
$$y(x - 2) = x + 3$$
$$y = \frac{x + 3}{x - 2}$$

The two equations given in Choice C are inverses.

Choice B: $y = 2(2x + 3) - 3 = 4x + 6$. The two given equations are **NOT** inverses.

Choice D: $y = \frac{(2x+1)-1}{2} = \frac{2x}{2} = x$ and $y = 2\left(\frac{x-1}{2}\right) + 1 = x - 1 + 1 = x$, so the two given equations are inverses.

31. B: The inequality that represents this scenario is $5n + 3s \geq 300$, where n is the number of non-student tickets that must be sold, and s is the number of student tickets that must be sold. The intercepts of this linear inequality can be found by substituting 0 for s and solving for n and then by substituting 0 for n and solving for s. The intercepts are $n = 60$ and $s = 100$. The solid line through the two intercepts represents the minimum number of each type of ticket that must be sold to offset production costs. The shading above the line represents sales which result in a profit for the school.

32. D: Set the first part of the analogy equal to y, then solve for x. When $y = \frac{1}{x}$, $x = \frac{1}{y}$. Consider choice D, when $y = x^2$, $x = \sqrt{y}$ for $y > 0$. On the other hand, when $y = x^2$, $x = \pm\sqrt{y}$; when $y = e^x$, $x = \ln y$ for $y > 0$; and when $y = x$, $x = y$.

33. C: Since 4 is the same as 2^2, $4^6 = (2^2)^6 = 2^{12}$. When dividing exponents with the same base, simply subtract the exponent in the denominator from the exponent in the numerator.

$$2^{12} \div 2^8 = 2^{12-8} = 2^4 = 16$$

Copyright © Mometrix Media. You have been licensed one copy of this document for personal use only. Any other reproduction or redistribution is strictly prohibited. All rights reserved.
This content is provided for test preparation purposes only and does not imply an endorsement by Mometrix of any particular political, scientific, or religious point of view.

Reading Comprehension

1. D: The correct answer is D because the author says that the ancient Britons were divided into many tribes and each had a king. The passage does say Choice A is incorrect because the passage says the ancient Britons had weapons, however, it doesn't say the weapons were the reason they fought. Choice B is incorrect because the text says the ancient Britons fought each other. It does not mention them fighting the Phoenicians. Choice C is incorrect because paragraph 6 says that there were no roads or bridges. The author gives this detail in order to describe the land. This detail does not support paragraph 7 which says that the ancient Britons regularly fought.

2. D: The correct answer is D because the word *grew* shows that the ancient Britons changed after the Phoenicians arrived. Choice A is incorrect because the answer choice only shows something that the Phoenicians did with the Britons; however, it does not describe how the Britons changed. Choice B is incorrect because basket-work is something the ancient Britons did before the Phœnicians interacted with them. Choice C is incorrect because it refers to the French and Belgians who moved to England.

3. B: The word *solitary* refers to isolation or being alone. Also, the sentence says that the islands are alone in a great expanse of water. So, this shows that other lands do not surround the islands. Choice A is incorrect because the sentence only talks about the size of the small islands around Scotland. It does not talk about where the islands are in relation to larger lands. Choice C is incorrect because it refers to people on the island, not the island in relation to the rest of the world. Choice D is incorrect because it shows the opposite of isolation; the sentence shows how the people who came to the islands (e.g., the French, Belgians, and Spanish) mixed with the ancient Britons.

4. B: The phrase "But seldom, if ever, ventured far from the shore" means that the boats didn't go far from shore. This implies that the boats may not have had much use in spite of the effort to build them. Choice A is incorrect because the text does not explain specifically how the boats were used. The author does describe weapons in the same paragraph. However, the phrase in the question focuses on the boats that the people made. The paragraph only moves onto weapons in the second sentence. Choice D is incorrect because the author does not connect the boats (discussed in paragraph 7) to the fortresses (discussed at the end of paragraph 6).

5. A: Choice A is the best answer because the first paragraph describes the location of the islands on the map. Choice B is incorrect because there were no roads that ran through England, Scotland, or Ireland during the time period that the passage is describing. Choice C is incorrect because the details about the small islands are only a small part of the first paragraph. Most of the paragraph describes how the islands appear on the map. Choice D is incorrect because the paragraph does not mention the size of the main islands. Also, the paragraph does not mention France or Belgium. Those countries are first mentioned in paragraph 4.

6. C: Choice C is the correct answer because this answer choice describes the lands by mentioning swamps and forests. Choice A is incorrect because it describes the coast, not the interior land of the island. Choice B is incorrect. It is about the interior of the island, but does not describe the land. Choice D is incorrect because it discusses the tribes and the people rather than the features of the lands (e.g., the swamps and forests).

7. B: The word *ardent* means passionate or devoted. Look at the word in the context of the sentence: "The dull precious metal seemed to flash with a reflection of her bright and <u>ardent</u> spirit," and think about Della's actions. Choice A is not correct because Della has planned and saved to buy her husband a Christmas present. So, she has not procrastinated (i.e., waited too long to start on

Copyright © Mometrix Media. You have been licensed one copy of this document for personal use only. Any other reproduction or redistribution is strictly prohibited. All rights reserved.
This content is provided for test preparation purposes only and does not imply an endorsement by Mometrix of any particular political, scientific, or religious point of view.

something). Choices C and D are synonyms. They are wrong because Della shows over and over again that she cares very much for her husband.

8. D: The first event in the story is Della preparing her hair (4). The next event in the story is Della preparing a meal for the night (2). Then, Della prays about the reaction that Jim will have to her hair (5). The next event is Jim assuring Della that he is not upset about her decision to cut her hair (3). Then, Jim gives the combs to Della (1). Finally, Della gives the fob chain to Jim (6).

9. C: The sentence "'I hunted all over town to find it.'" is an exaggeration or hyperbole. The other answer choices have similes.

10. C: In the beginning of the story, Della has her hair cut off. She does this so she can get money for Jim's gift. However, she is worried about how she will appear to her husband and is concerned that he will be very angry or worse. When Jim keeps asking about Della's hair, it almost seems as if he is angry about it.

11. A: The entire story is about how Della and Jim are deeply in love. They sell a prized possession to buy a present for the other. Choice B is wrong because the overall meaning of this passage is not about a holiday. Choice C is wrong because it fails to recognize how Della and Jim overcome difficulty and have a deeper love for each other after giving their gifts. Choice D is wrong because the focus of the passage is not on buying the right gift for a loved one. Instead, the focus is on giving up a thing that you care about to show how much you love someone.

12. D: The narrator of the passage is not named. Della and Jim are simply characters in the story. O. Henry is the author of the story, not the narrator.

13. C: The question asks which job is less desirable than homesteading according to the writer. Choice C is correct because Stewart speaks of going out to wash as less preferable than homesteading. Choice A is incorrect because the letter does not mention cooking or restaurants. Choice B is also incorrect. The reason is that there is no mention of opening a bed and breakfast in the letter. Choice D is also wrong because the letter shares nothing about teaching.

14. D: Growing potatoes is so simple that her six-year-old can do so with little help. The issue is not child labor. So, choice A is incorrect. Stewart does not mention schooling. So, this makes choice B incorrect. Stewart's point has nothing to do with whether women work as hard as men. So, choice C is not the correct response.

15. B: Stewart mentions the hard work of laundry, but she does not speak of any enjoyment of it as she does about homesteading. Choice A cannot be correct because Stewart does not seem to be lazy. Choice C is possible, but there is no evidence to support that idea. Choice D cannot be correct because no reason for anger is given.

16. D: Stewart says that homesteading is a lonely task. For example, she mentions that "persons afraid of ...loneliness had better let ranching alone." Stewart explains in the letter that her work uses less strength than washing. So, choice A is wrong. Choice B is also incorrect because it is addressed in the first paragraph of the letter. Choice C is incorrect because it is mentioned as an advantage in the first paragraph.

17. A: Stewart directly states that women would have "the added satisfaction of knowing that their job will not be lost to them if they care to keep it." Although going on strike was common at the time, Stewart does not mention it. So, choice B is incorrect. Denver is in the Rocky Mountains, but

Copyright © Mometrix Media. You have been licensed one copy of this document for personal use only. Any other reproduction or redistribution is strictly prohibited. All rights reserved.
This content is provided for test preparation purposes only and does not imply an endorsement by Mometrix of any particular political, scientific, or religious point of view.

landslides are not mentioned as a risk. So, choice C cannot be the correct answer. Fire is always a risk, but Stewart does not bring it up in the letter. So, choice D is not the correct answer.

18. C: The letter is very positive and full of reasons on why homesteading is a good choice. Choice A is incorrect. The reason is that there is no complaint of hard work, weather, or loneliness. Choice B is also not correct because the letter does not speak of sadness or loneliness. Instead, Stewart rejoices in the good success of the homestead. Also, choice D is not correct because there is no hopelessness expressed in the letter.

19. C: The professor argues that small, woody material is left on the tops of trees and is less likely to be reached by fire.

20. D: Choice D is not a supporting detail because it is a definition of salvage logging. The other choices are supporting details of the forest manager's argument.

21. B: Plot A was salvage logged and burned worse than the unmanaged plot (Plot B). This study supports the professor's view that salvage logging increases the risk and severity of fire.

22. A: The question asks which option is the chief argument regarding fire prevention. Choices B and D are not helpful for fire prevention. Choice C is incorrect because logging decreases soil temperature. Choice B is a supporting detail from the passage but is not the main idea. Choice C contradicts the passage. Choice D is not mentioned in the passage.

23. D: The professor says the larger trees in old growth forests are more resistant to fire than smaller, younger trees. Choices A, B, and C all agree with the professor.

24. C: Both the manager and the professor discuss the importance of seedling growth after a fire. So, a study looking at the regrowth of seedlings in logged and unmanaged forests would potentially provide support for both arguments (as well as possibly showing problems with both arguments).

25. C: The first section covers how people want to change and the options that are available to them. As people move towards a healthier lifestyle, they can work at small steps on their way to better health. Choice A is incorrect because the first section does not mention menus. Choice B is incorrect because the paragraph lists benefits of certain foods but does not suggest any are better than others. Choice D is incorrect because the passage mentions that the process is a struggle for many, not an impossible task.

26. D: The overall passage is information for someone who is interested in the food industry. So, choice A is not correct. The reason is that there is not much of a comparison or contrast between the Food Pyramid and MyPlate. Choice B is incorrect because the focus of the passage is not on government involvement. Choice C is incorrect because the text says some groups disagree about certain dietary suggestions but does not make recommendations.

27. A: Choice B is wrong because there is no mention of Harvard's guidelines replacing MyPlate. Choice C is also incorrect. It is another approach to dietary guidelines, but it was included to support a claim, not provide diet options. Choice D is wrong because Harvard is included for their material, not their Ivy League status.

28. C: This passage was written for people who have an interest in the food industry. So, choice A is wrong because the passage is not meant for those who are new to dieting. Choice B is wrong because experienced professionals don't need introductory-level material on preparing food. Choice D is incorrect because there is no mention of parents or children in the passage.

Copyright © Mometrix Media. You have been licensed one copy of this document for personal use only. Any other reproduction or redistribution is strictly prohibited. All rights reserved.
This content is provided for test preparation purposes only and does not imply an endorsement by Mometrix of any particular political, scientific, or religious point of view.

29. B: Each section of the passage has the tone of educating readers. So, choice A is wrong because the tone is not blaming readers. Choice C is close, but it is not the best answer choice. Choice D is wrong because the tone of the material is not snobbish or cocky.

30. C: The Constitution overrides state law in some cases. Answer choices B and D are quite obviously factually incorrect. The last choice, A, is also not stated or implied in the passage, but this answer might seem correct to someone who hadn't read the passage carefully.

31. A: Although state courts exercised judicial review prior to the ratification of the Constitution, the doctrine is most often traced to the landmark US Supreme Court decision Marbury v. Madison (1803), which struck down an act of Congress as unconstitutional. This sentence means that laws passed by Congress must comply with the Constitution. None of the other choices do.

32. A: This is an informational passage about history, so *historical* could be used to describe its contents. None of the other words fit this passage. *Introspective* refers to a passage that describes a persons internal thoughts or considerations. *Prospective* refers to the future. And *demonstrative* refers to expressing feeling or emotion.

33. A: When the 13 colonies ratified the Constitution, they endorsed it.

34. D: When Oliver Wendell Holmes dissented from many opinions, it meant he disagreed with the majority on the court. So his opinion was *different.*

35. C: As the passage explains, the Supreme Court's ruling in the 1803 case of *Marbury v. Madison,* which invalidated a law passed by Congress on the grounds that it violated the Constitution, is the foundation of the legal concept of judicial review.

36. C: According to the passage, after a period in which the Supreme Court struck down over 200 laws as unconstitutional, FDR threatened to "pack" the Court, which led to fewer laws being struck down. This is in reference to FDR's plan to add additional justices to the Supreme Court, who would have probably have been supporters of his who could be counted on to uphold the constitutionality of laws FDR got Congress to pass.

Copyright © Mometrix Media. You have been licensed one copy of this document for personal use only. Any other reproduction or redistribution is strictly prohibited. All rights reserved.
This content is provided for test preparation purposes only and does not imply an endorsement by Mometrix of any particular political, scientific, or religious point of view.

IBEW Practice Test #2

Algebra and Functions

1. Kendra uses the pie chart below to represent the allocation of her annual income. Her annual income is $40,000.

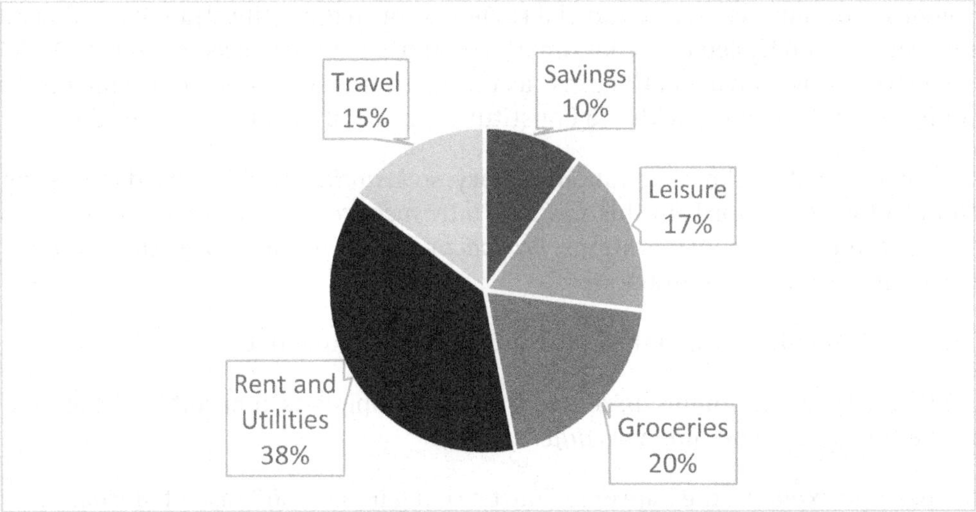

Which of the following statements is true?

a. The amount of money she spends on travel and savings is more than $11,000.
b. The amount of money she spends on rent and utilities is approximately $15,000.
c. The amount of money she spends on groceries and savings is more than $13,000.
d. The amount of money she spends on leisure is less than $5,000.

2. Which of the following expressions is equal to x^3x^5?

a. x^2
b. x^8
c. $2x^8$
d. x^{15}

3. Which of these statements is true for functions $f(x)$, $g(x)$, and $h(x)$?

$$f(x) = -2x + 1$$
$$g(x) = -2x^2 + 1$$
$$h(x) = -2x^3 + 1$$

a. The degree of each polynomial function is –2.
b. The leading coefficient of each function is –2.
c. Each function has exactly one real zero.
d. Each function has an x-intercept at (1,0).

Copyright © Mometrix Media. You have been licensed one copy of this document for personal use only. Any other reproduction or redistribution is strictly prohibited. All rights reserved.
This content is provided for test preparation purposes only and does not imply an endorsement by Mometrix of any particular political, scientific, or religious point of view.

4. Which of these is equivalent to $\log_y 64$ if $2\log_8 y + \log_8 4 = 2$?

a. 1
b. 2
c. 3
d. 6

5. Which of the following statements is (are) true when $f(x) = \frac{x^2-x-6}{x^3+2x^2-x-2}$?

I. The graph $f(x)$ has vertical asymptotes at $x = -2$, $x = -1$, and $x = 1$.
II. The x- and y-intercepts of the graph of $f(x)$ are both 3.

a. I only
b. II only
c. I and II
d. Neither statement is true.

6. Complete the analogy: x^3 is to $\sqrt[3]{y}$ as ...

a. $x + a$ is to $x - y$.
b. e^x is to $\ln y$, $y > 0$.
c. $\frac{1}{x}$ is to y, $x, y \neq 0$.
d. $\sin x$ is to $\cos y$.

7. The equations $y = 8x - 4$ and __________ will form parallel lines when graphed.

a. $y = \frac{1}{8}x - 4$
b. $y = 2x - 1$
c. $y = \frac{1}{8}x - 4$
d. $y = 8x - 1$

8. If $f(x)$ and $g(x)$ are inverse functions, which of these is the value of x when $f(g(x)) = \frac{1}{8}$?

a. -8
b. $\frac{1}{8}$
c. 8
d. $-\frac{1}{8}$

Copyright © Mometrix Media. You have been licensed one copy of this document for personal use only. Any other reproduction or redistribution is strictly prohibited. All rights reserved.
This content is provided for test preparation purposes only and does not imply an endorsement by Mometrix of any particular political, scientific, or religious point of view.

9. It rained every day last week except Friday. If it rained twice as much on Monday as it did on Thursday, which of the following graphs could model the week's rainfall in inches?

a.
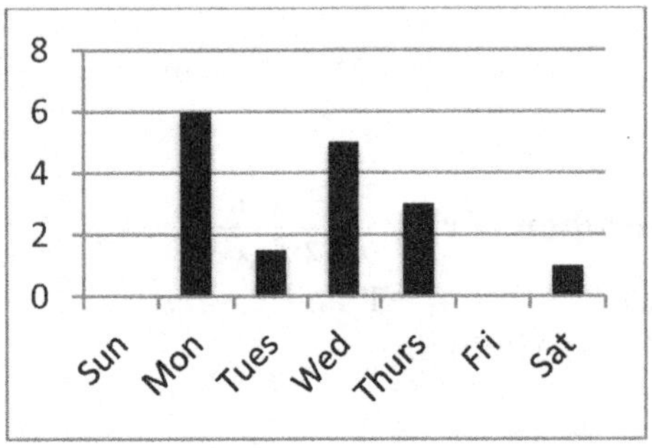

b.
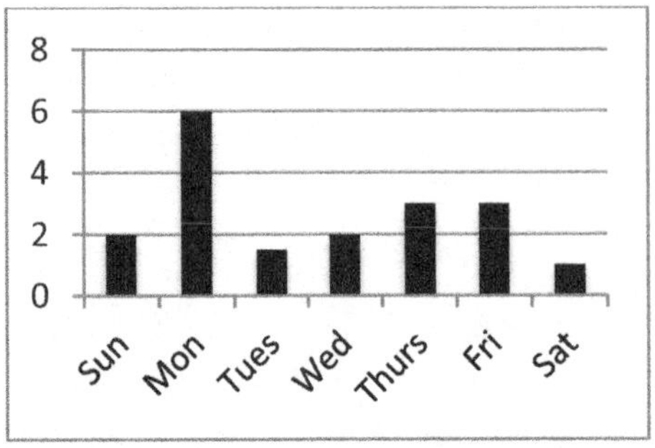

c.
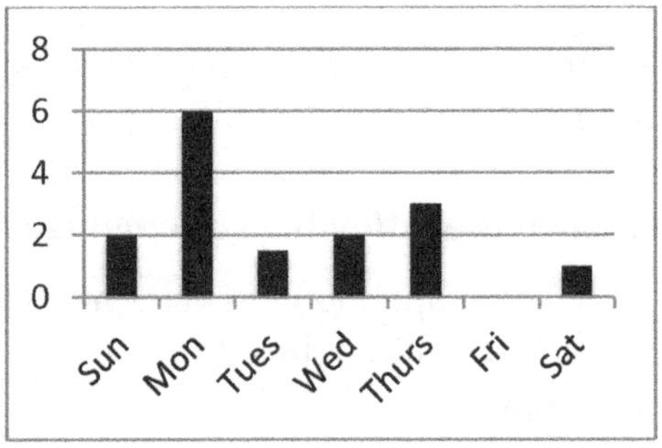

d.
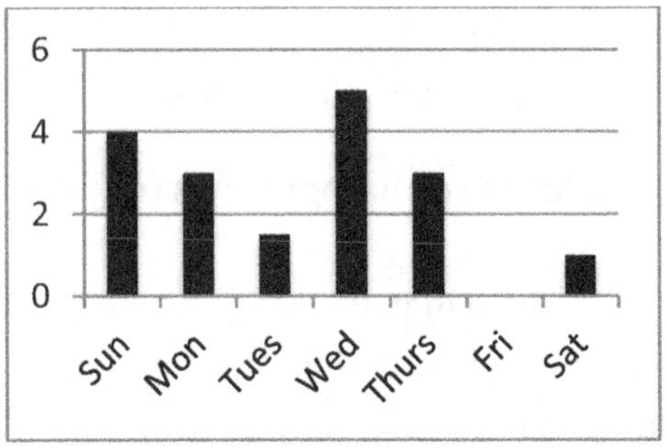

10. Expand the following: $9x(3x^2 + 2x - 9)$

a. $27x^2 + 18x - 81$
b. $27x^3 + 18x^2 - 81x$
c. $12x^3 + 11x^2 - x$
d. $27x^3 + 18x^2 - 18x$

11. Which of the following expressions is equivalent to the equation $3x^2 + 4x - 15$?

a. $(x - 3)(x + 5)$
b. $(x + 5)(3 + x^2)$
c. $x(3x^2 + 4 - 15)$
d. $(3x - 5)(x + 3)$

12. The equations $y = 0.5x + 7.4$ and __________ will form perpendicular lines when graphed.

a. $y = 2x + 3.4$
b. $y = -2x + 5.5$
c. $y = 0.5x - 1$
d. $y = -0.5x - 3.6$

Copyright © Mometrix Media. You have been licensed one copy of this document for personal use only. Any other reproduction or redistribution is strictly prohibited. All rights reserved.
This content is provided for test preparation purposes only and does not imply an endorsement by Mometrix of any particular political, scientific, or religious point of view.

Refer to the following for questions 13–14:

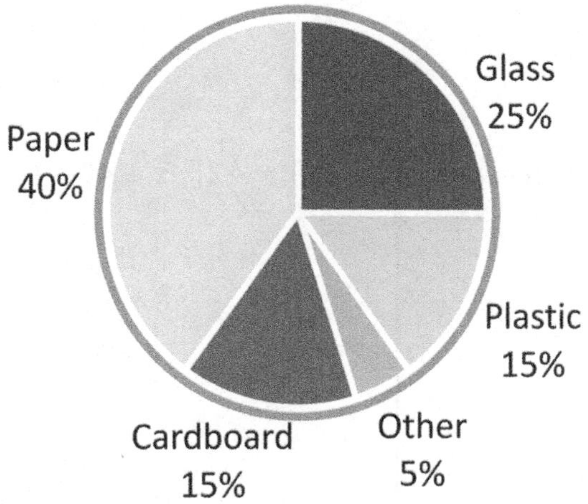

13. The Charleston Recycling Company collects 50,000 tons of recyclable material every month. The chart shows the kinds of materials that are collected by the company's five trucks. What is the second most common material that is recycled?

a. Cardboard
b. Glass
c. Paper
d. Plastic

14. If 50,000 tons of material are recycled every month, about how much paper is recycled every month?

a. 40,000 tons
b. 50,000 tons
c. 20,000 tons
d. 15,000 tons

15. What equation is graphed below?

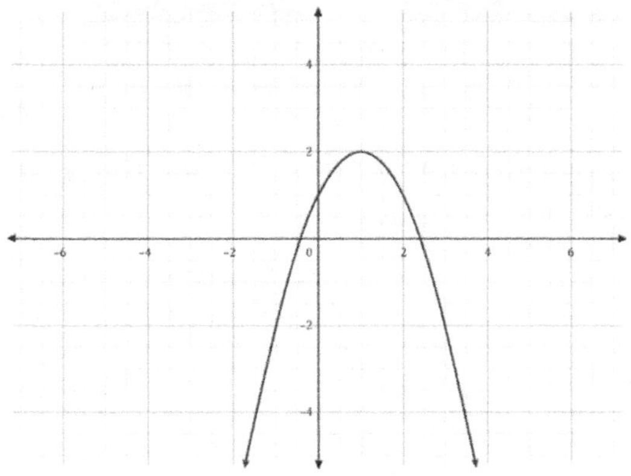

a. $y = -2x^2 - 2x + 1$
b. $y = -x^2 + 2x + 1$
c. $y = -x^2 - 2x + 2$
d. $y = -\frac{1}{2}x^2 - x + 2$

Copyright © Mometrix Media. You have been licensed one copy of this document for personal use only. Any other reproduction or redistribution is strictly prohibited. All rights reserved.
This content is provided for test preparation purposes only and does not imply an endorsement by Mometrix of any particular political, scientific, or religious point of view.

16. Factor the following expression: $x^2 + x - 12$

a. $(x - 2)(x + 6)$
b. $(x + 6)(x - 2)$
c. $(x - 4)(x + 3)$
d. $(x + 4)(x - 3)$

17. Which of the graphs shown represents $f(x) = -2|-x + 4| - 1$?

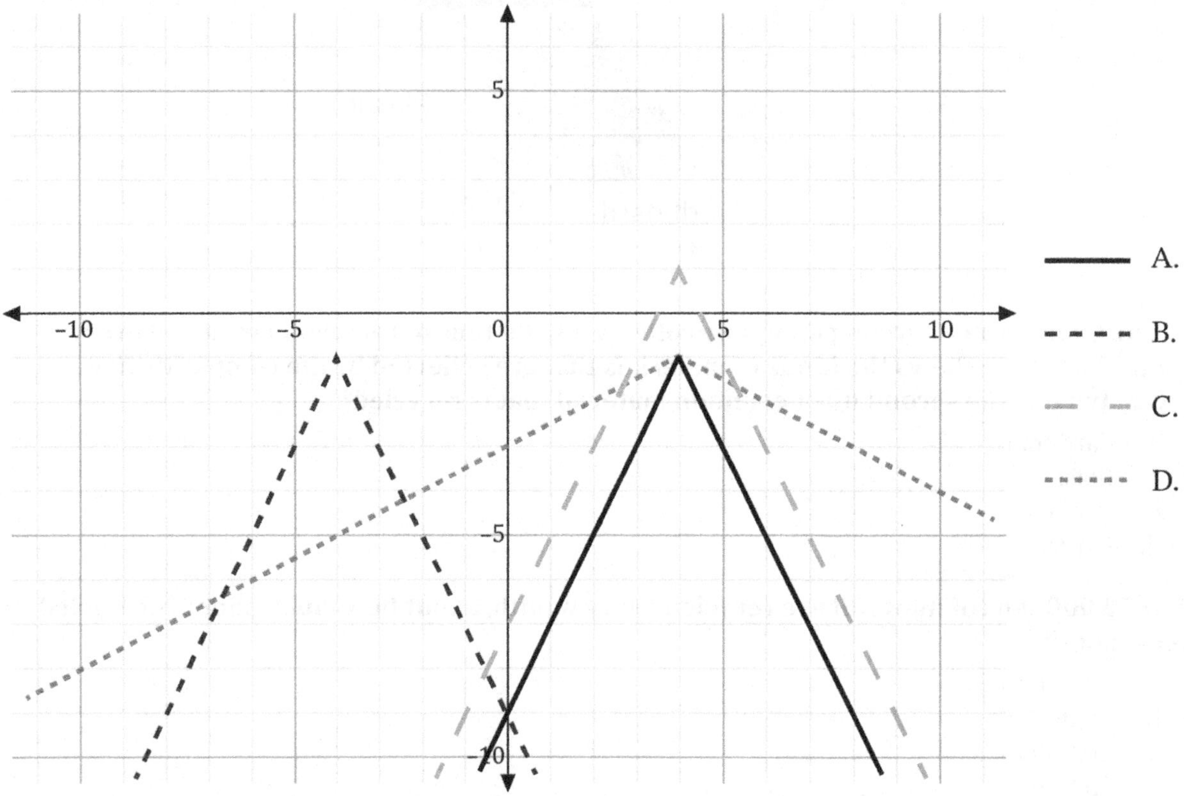

a.
b.
c.
d.

18. Suppose Line 1 has a slope of -5. In order graph a line that is perpendicular to Line 1, the slope of the new line will need to be ___________.

a. $\frac{1}{5}$
b. -5
c. 5
d. $-\frac{1}{5}$

19. If x and y are positive integers, which of the following expressions is equivalent to $(xy)^{7y} - (xy)^{y}$?

a. $(xy)^{6y}$
b. $(xy)^{7y-1}$
c. $(xy)^{y}[(xy)^{7} - 1]$
d. $(xy)^{y}[(xy)^{6y} - 1]$

Copyright © Mometrix Media. You have been licensed one copy of this document for personal use only. Any other reproduction or redistribution is strictly prohibited. All rights reserved.
This content is provided for test preparation purposes only and does not imply an endorsement by Mometrix of any particular political, scientific, or religious point of view.

20. The table below shows the cost of renting a bicycle for 1, 2, or 3 hours. Which answer choice shows the equation that best represents the data? Let C represent the cost of the rental and h stand for the number of hours of rental time.

Hours	1	2	3
Cost	$3.60	$7.20	$10.80

a. $C = 3.60h$
b. $C = h + 3.60$
c. $C = 3.60h + 10.80$
d. $C = \frac{10.80}{h}$

21. Lauren had $80 in her savings account. When she received her paycheck, she put some money in her savings account. This brought the balance up to $120. By what percentage did the total amount in her account increase by putting this amount in her savings account?

a. 35%
b. 40%
c. 50%
d. 80%

22. The equation below calculates the growth of an apple tree, where h is the height in feet and y is the number of years since the tree was planted. How many inches does the tree grow each year?

$$h = 2.5 + 0.75y$$

a. 0.75
b. 2.5
c. 9
d. 30

23. If $a = 4$, $b = 3$, and $c = 1$, then what is the value of $\frac{a(b-c)}{b(a+b+c)}$?

a. $\frac{4}{13}$
b. $\frac{1}{3}$
c. $\frac{1}{4}$
d. $\frac{1}{6}$

24. What is the value of b when $f(g(b)) = -6$, if $f(x)$ and $g(x)$ are inverse functions?

a. -6
b. $-\frac{1}{6}$
c. $\frac{1}{6}$
d. 6

Copyright © Mometrix Media. You have been licensed one copy of this document for personal use only. Any other reproduction or redistribution is strictly prohibited. All rights reserved.
This content is provided for test preparation purposes only and does not imply an endorsement by Mometrix of any particular political, scientific, or religious point of view.

25. Which of these is the equation graphed below?

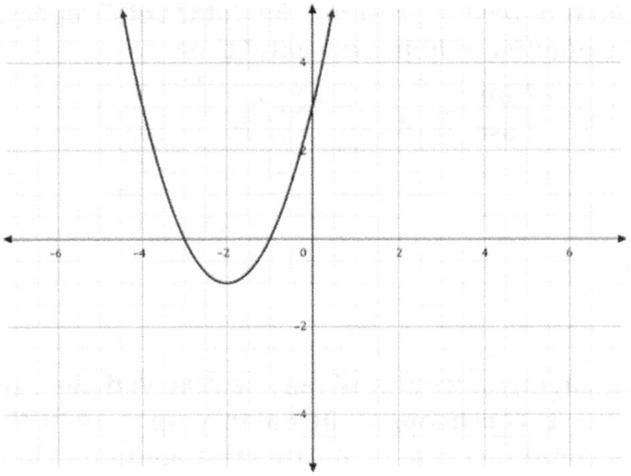

a. $y = x^2 + 4x + 3$
b. $y = x^2 - 2x - 1$
c. $y = 2x^2 + 4x + 1$
d. $y = \frac{1}{2}x^2 - 2x + 3$

26. Complete the analogy: $y = \frac{1}{\sqrt{x}}$ is to $x = y^{-2}$ as ...

a. $y = 2x$ is to $x = y^2$.
b. $y = e^x$ is to $x = \frac{1}{e^y}$, where $y > 0$.
c. $y = \frac{1}{x^2}$ is to $x = y^2$, where $x, y \neq 0$.
d. $y = \cos x$ is to $x = \cos^{-1} y$.

27. Solve $2x^3y^2 - 12x^2y = 4x^3y - 6x^2y^2$ for x and y.

a. $x = \{0, -3\}, y = \{0\}$
b. $x = \{0\}, y = \{0\}$
c. $x = \{0, -3\}, y = \{0,2\}$
d. $x = \{0\}, y = \{0, -2\}$

28. Solve the system of equations: $\begin{cases} 3x + y = 1 \\ -4x + 3y = 29 \end{cases}$.

a. $(-2, 7)$
b. $(-1, 3)$
c. $\left(0, \frac{1}{2}\right)$
d. $\left(2, -\frac{5}{2}\right)$

Copyright © Mometrix Media. You have been licensed one copy of this document for personal use only. Any other reproduction or redistribution is strictly prohibited. All rights reserved.
This content is provided for test preparation purposes only and does not imply an endorsement by Mometrix of any particular political, scientific, or religious point of view.

29. Ben is purchasing notebooks and boxes of pencils for a class he is teaching. Notebooks cost \$1 each and pencils cost \$3 per box. If the x-axis represents boxes of pencils, the y-axis represents notebooks, and Ben can spend a maximum of \$30, which of the following graphs shows the number of notebooks and pencils Ben can purchase?

a.
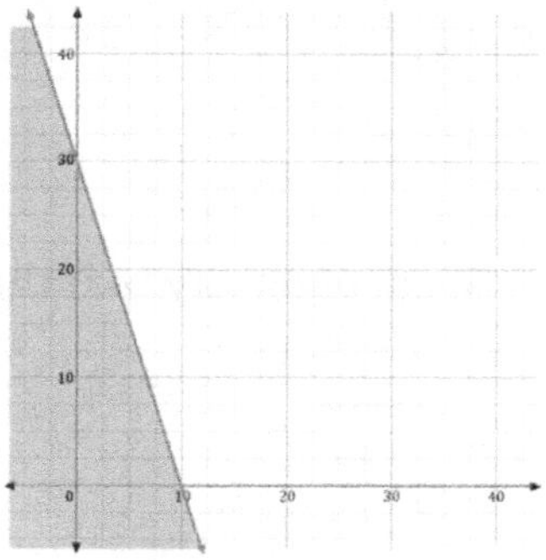

c.
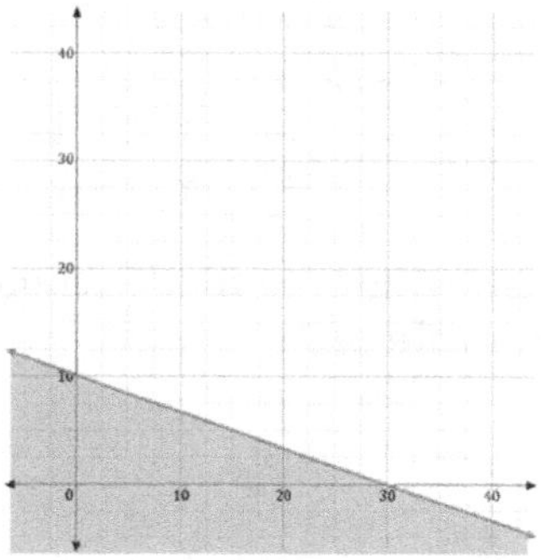

b.
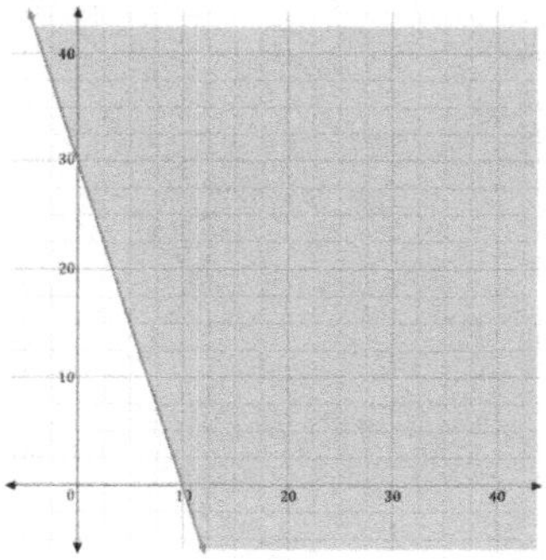

d.
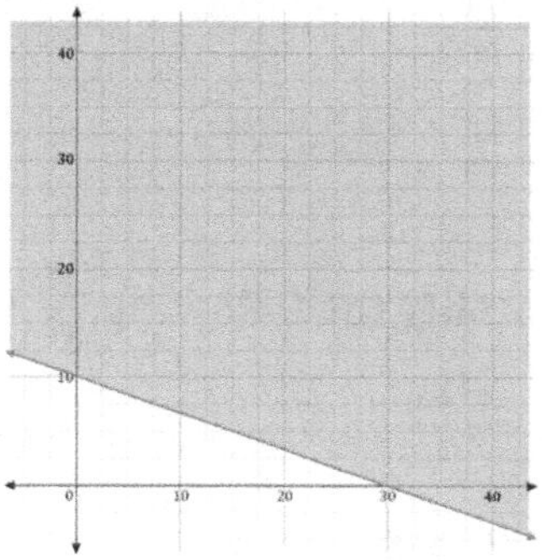

30. Which of these equations is represented by the graph below?

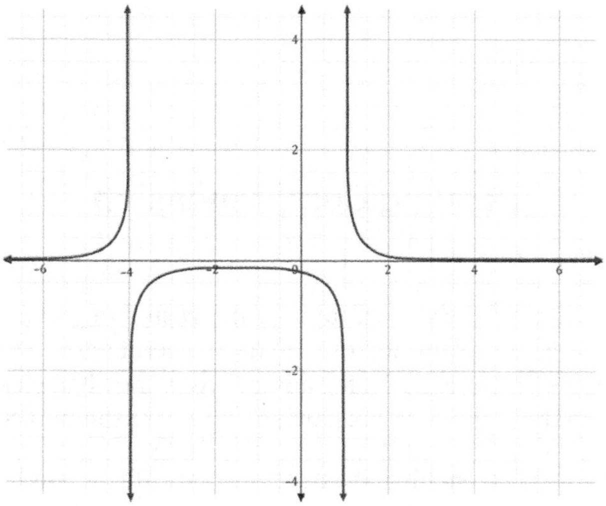

a. $y = \frac{2x-2}{x^2+3x-4}$

b. $y = \frac{1}{x^2+3x-4}$

c. $y = \frac{1}{x+4} + \frac{1}{x-1}$

d. $y = \frac{2}{3x^2+9x-12}$

Copyright © Mometrix Media. You have been licensed one copy of this document for personal use only. Any other reproduction or redistribution is strictly prohibited. All rights reserved.
This content is provided for test preparation purposes only and does not imply an endorsement by Mometrix of any particular political, scientific, or religious point of view.

31. Which of these functions includes 1 as an element of the domain and 2 as an element of the range?

a. $y = \frac{1}{x-1} + 1$

b. $y = -\sqrt{x+2} - 1$

c. $y = |x+2| - 3$

d. $y = \begin{cases} x & x < -1 \\ -x-3 & x \geq -1 \end{cases}$

32. A two-digit number is chosen at random. What is the probability that the chosen number is a multiple of 7?

a. $\frac{1}{10}$

b. $\frac{1}{9}$

c. $\frac{11}{90}$

d. $\frac{13}{90}$

33. Simplify the following: $\frac{x^2}{y^2} + \frac{x}{y^3}$

a. $\frac{x^3+x}{y^3}$

b. $\frac{x^2+xy}{y^3}$

c. $\frac{x^2y+xy}{y^3}$

d. $\frac{x^2y+x}{y^3}$

Reading Comprehension

Refer to the following for questions 1–6:

George Washington Carver was always interested in plants. When he was a child, he was known as the "plant doctor." He had a secret garden where he grew all kinds of plants. People would ask him for advice when they had sick plants. Sometimes he'd take their plants to his garden and nurse them back to health.

Later, when he was teaching at Tuskegee Institute, he put his plant skills to good use. Many people in the South had been growing only cotton on their land. Cotton plants use most of the nutrients in the soil. (Nutrients provide nourishment

Copyright © Mometrix Media. You have been licensed one copy of this document for personal use only. Any other reproduction or redistribution is strictly prohibited. All rights reserved.
This content is provided for test preparation purposes only and does not imply an endorsement by Mometrix of any particular political, scientific, or religious point of view.

to plants.) So the soil becomes "worn out" after a few years. Eventually, cotton will no longer grow on this land.

This was especially bad for poor African American farmers, who relied on selling cotton to support themselves. Carver was dedicated to helping those farmers, so he came up with a plan.

Carver knew that certain plants put nutrients back into the soil. One of those plants is the peanut! Peanuts are also a source of protein.

Carver thought that if those farmers planted peanuts, the plants would help restore their soil, provide food for their animals, and provide protein for their families—quite a plant! In 1896 peanuts were not even recognized as a crop in the United States, but Carver would help change that.

Carver told farmers to rotate their crops: plant cotton one year, then the next year plant peanuts and other soil-restoring plants, like peas and sweet potatoes. It worked! The peanut plants grew and produced lots of peanuts. The plants added enough nutrients to the soil so cotton grew the next year.

1. Why was George Washington Carver known as the "plant doctor"?

a. He grew peanuts on sick soil.
b. He was a plant pathologist.
c. He could nurse sick plants back to health.
d. He knew plants could put nutrients back into the soil.

2. How is this passage structured?

a. cause and effect
b. problem and solution
c. chronological order
d. proposition and support

3. According to the passage, what problem were cotton farmers facing?

a. They needed food for their animals.
b. Peanuts were not recognized as a crop in the United States.
c. They were growing too much cotton.
d. The cotton had stripped the land of its nutrients.

4. How did Carver resolve the farmers' problem?

a. He told the farmers to rotate their crops.
b. He came up with a plan.
c. He invented the peanut.
d. He gave advice to farmers with sick plants.

5. This passage is mainly about

a. how George Washington Carver became a teacher at the Tuskegee Institute.
b. how George Washington Carver helped farmers improve their crop production.
c. why George Washington Carver studied plants.
d. how George Washington Carver made peanuts a recognized crop in the United States.

Copyright © Mometrix Media. You have been licensed one copy of this document for personal use only. Any other reproduction or redistribution is strictly prohibited. All rights reserved.
This content is provided for test preparation purposes only and does not imply an endorsement by Mometrix of any particular political, scientific, or religious point of view.

6. According to this passage, "crop rotation" can be described as

a. Planting a soil-depleting crop like cotton one year and then planting a soil-restoring plant like peanuts the next year
b. Growing only one crop on the land year after year
c. Nursing sick plants back to health
d. Studying plants at an institute

Refer to the following for questions 7–11:

JOB HAZARDS ANALYSIS

1.0 PURPOSE AND SCOPE

This procedure describes the Job Hazard Analysis (JHA) process for identifying, evaluating, controlling, and communicating potential hazards and environmental impacts associated with operations or work by the Tank Operations Contractor (TOC). It applies to all TOC work activities, including the performance of field work involving general plant maintenance, operations, and environmental remediation. This procedure applies to subcontractors who do not have an approved job hazard analysis process. Everyone is required to work safely and to maintain a safe work environment. Training procedures have been reviewed to ensure that workers are trained to the general hazards associated with work at the tank farms. Visitors should be briefed on the general safety hazards they may be exposed to and controls expected of them as part of their orientation.

2.0 IMPLEMENTATION

This procedure is effective on the date shown in the header.

3.0 RESPONSIBILITIES

Responsibilities are contained within Section 4.0.

4.0 Methods for Implementation of Controls

In order to effectively implement necessary controls to mitigate or eliminate hazards to the workers, the following guidelines should be used:

4.1. The following hierarchy of methods to eliminate or mitigate hazards shall be used in descending order, when feasible and appropriate:

1. Eliminate the hazard or substitution (e.g., different chemical cleaning agent)
2. Utilize engineering controls (e.g., ventilation)
3. Administrative controls (e.g., dose monitoring)
4. Personal protective equipment (PPE) (e.g., self-contained breathing apparatus)

4.2. Controls within the qualification or training of the worker that are often used do not need to be discussed in the work instructions. Examples: Use of leather gloves, safety glasses of the proper type that the worker normally uses.

4.3. Controls within the qualification or training of the worker that are seldom used, and are applicable to the entire work activity, should be placed in the precautions as a reminder that the hazard exists and the workers are expected to take the appropriate actions. Examples: Use of

Copyright © Mometrix Media. You have been licensed one copy of this document for personal use only. Any other reproduction or redistribution is strictly prohibited. All rights reserved.
This content is provided for test preparation purposes only and does not imply an endorsement by Mometrix of any particular political, scientific, or religious point of view.

hearing protection due to a noisy environment at the job site, or observation of overhead lines when they are present at the job site.

4.4. Controls within the qualification and training of the workers, but for hazards that are introduced at specific steps or by specific actions during the job, should have a warning or caution statement immediately prior to the step but require no detailed mitigation instructions in the work instructions. Example: a warning for the release of pressure when breaching a system that may have residual pressure.

4.5. Controls not within the qualification and training of the workers for hazards should have detailed instructions for how the workers are to mitigate the hazard and should be in the work instructions or procedure in a way that is prominent and prevents or mitigates the hazard. Example: the steps required to successfully release the pressure on a system in an operation which is not normally performed.

7. This document is what type of document?

a. A government request for proposals
b. A process for making rules for working safely
c. A portion of a contract
d. A set of rules for working efficiently

8. According to the procedure, if a worker is exposed to a hazardous chemical, which of the following is the last thing that should be tried to prevent injury or illness?

a. Use a different chemical.
b. Install fans to keep fumes away from the worker.
c. Measure the amount of exposure of each affected worker.
d. Give the worker protective clothing.

9. Welders must always use goggles and are taught to use them as part of their basic training. According to the text, the use of goggles during specific welding operations should

a. be prominently displayed at the beginning of the work instruction.
b. be displayed as a caution prior to the welding step described in the work instruction.
c. be described in detail in the work instruction.
d. not be discussed in the work instruction.

10. This passage would normally need to be read and understood by

a. managers at the site.
b. laborers at the site.
c. visitors to the site.
d. workers making deliveries to the site.

11. Which of the following requires the most comprehensive description within the work instructions?

a. Controls that are part of the worker's training and are used routinely
b. Controls that are part of the worker's training but that are seldom used
c. Controls that are part of the worker's training and that are required for specific steps in the work procedure
d. Controls that are not part of the worker's training

Copyright © Mometrix Media. You have been licensed one copy of this document for personal use only. Any other reproduction or redistribution is strictly prohibited. All rights reserved.
This content is provided for test preparation purposes only and does not imply an endorsement by Mometrix of any particular political, scientific, or religious point of view.

Refer to the following for questions 12–21:

This passage is a re-telling of a traditional American Indian legend.

The Black Crow.

In ancient times, the people hunted the buffalo on the Great Plains. These huge animals were their source of food and clothing. With stone-tipped spears, they stalked the great beasts through the tall grasses. It was difficult and dangerous work, but they were forced to do it in order to survive.

At that time, there were many crows flying above the plains, as there are today. But unlike the crows we see now, these birds were white. And they were friends to the buffalo, which caused the hunters no end of travail. The white crows flew high above the plains, where they could see all that was happening below. And when they saw that hunters were approaching the herd, they would warn the buffalo. Swooping down low, they would land on the heads of the great beasts and call out to them: "Beware! Beware! Hunters are coming from the south! Caw, caw. Beware!" And the buffalo would stampede, leaving the hunters empty-handed.

This went on for some time, until the people were hungry and something needed to be done. A council was convened, and the chief of the people spoke to them. "We must capture the chief of the crows and teach him a lesson," he said. "If we can frighten him, he will stop warning the buffalo when our hunters approach, and the other crows will stop as well."

The old chief then brought out a buffalo skin, one with the head and horns still attached. "With this, we can capture the chief of the crows," he said. And he gave the skin to one of the tribe's young braves, a man known as Long Arrow. "Disguise yourself with this, and hide among the buffalo in the herd," the chief told Long Arrow. "Then, when the chief of the crows approaches, you will capture him and bring him back to the tribe."

So Long Arrow donned the buffalo skin disguise and went out onto the plains. Carefully, he approached a large herd of buffalo and mingled among them, pretending to graze upon the grasses. He moved slowly with the herd as they sought fresh food, and he waited for the great white bird that was the chief of the crows.

The other braves made ready for the hunt. They prepared their stone-tipped spears and arrows, and they approached the grazing herd of beasts, hiding in ravines and behind rocks to try to sneak up on them. But the crows, flying high in the sky, saw everything. The chief of the crows saw the men in the ravines and tall grasses, and eventually he came gliding down to warn the buffalo of the approaching hunters.

Hearing the great white crow's warning, the herd ran from the hunters. All stampeded across the plains except Long Arrow, still in his disguise. Seeing that Long Arrow remained, and thinking that he was a buffalo like all the others, the great white crow flew to him and landed upon his head. "Caw, caw. Hunters are approaching! Have you not heard my warning? Why do you remain here?" But as the great bird cried out, Long Arrow reached from under his disguise and grabbed the bird's feet, capturing him. He pushed him into a rawhide bag and brought him back to the tribal council.

The people debated what to do with the chief of the crows. Some wanted to cut his wings so that he could not fly. Some wanted to kill him, and some wanted to remove his feathers as punishment for making the tribe go hungry. Finally, one brave strode forward in anger, grabbed the rawhide bag that held the bird, and before anyone could prevent it, threw it into the fire.

Copyright © Mometrix Media. You have been licensed one copy of this document for personal use only. Any other reproduction or redistribution is strictly prohibited. All rights reserved.
This content is provided for test preparation purposes only and does not imply an endorsement by Mometrix of any particular political, scientific, or religious point of view.

As the fire burned the rawhide bag, the big bird struggled to escape. Finally, he succeeded in getting out of the bag and managed to fly out of the fire, but his feathers were singed and covered with black soot from the fire. The chief of the crows was no longer white; he was black – as crows are today.

And from that day forward, all crows have been black. And although they fly above the plains and can see all that transpires below, they no longer warn the buffalo that hunters are approaching.

12. According to the passage, the people used stone spears to hunt the buffalo because

a. They had no metal.
b. They had no horses.
c. They needed to eat.
d. They were plentiful.

13. The word *travail* (Line 14) means

a. Travel
b. Difficulty
c. Anger
d. Fear

14. Which statement best describes what the chief of the crows represents in this passage?

a. He symbolizes all that is evil.
b. He is a symbol representing all crows.
c. He represents the animal kingdom.
d. He represents other predators who compete with the tribe.

15. Which of the following best describes the people's motivation for wanting to capture the chief of the crows?

a. They hated birds.
b. They wanted to turn him black.
c. They wanted to eat him.
d. They were hungry.

16. Long Arrow's activities among the herd while disguised imply that he

a. Had time to kill.
b. Wanted to fool the buffalo.
c. Wanted to fool the crows.
d. Had forgotten his stone-tipped spear.

17. In this tale, the rawhide bag and stone-tipped spears are both details that

a. Are important for the outcome of the tale
b. Paint a picture of the culture of the people
c. Make it clear that the people were dependent upon the buffalo
d. Show how the people hunted

Copyright © Mometrix Media. You have been licensed one copy of this document for personal use only. Any other reproduction or redistribution is strictly prohibited. All rights reserved.
This content is provided for test preparation purposes only and does not imply an endorsement by Mometrix of any particular political, scientific, or religious point of view.

18. Why might the chief of the crows have landed upon Long Arrow's head after seeing the other buffalo stampede away?

a. He thought his warning had not been heard.
b. He wanted to see the disguise.
c. He thought that Long Arrow was an injured buffalo.
d. He had no fear of men.

19. Once the bird has been caught, what emotions are revealed by the people's deliberations about how to deal with him?

a. Anger
b. A calm resolve to change the birds' behavior
c. A feeling of celebration now that the bird has been caught
d. Hunger

20. What does the story tell us about why Long Arrow was selected for this task?

a. He was the bravest man in the tribe.
b. He was related to the chief.
c. He was able to act like a buffalo.
d. The story says nothing about why he was selected.

21. What does this story suggest that the American Indians thought of crows?

a. They were dirty animals.
b. They were clever animals.
c. They were selfish animals.
d. They disliked the people in the tribe.

Refer to the following for questions 22–28:

Zakov arrived at the prison shortly before noon, having dropped Ludmilla off at the train station beforehand. He was relieved to find that Gorkhi had gone, for he found the chief to be a bit bombastic, and was disposed to conduct the interrogation without him if he could. He was fortunate in that one of Gorkhi's men recognized him and let him in to see the prisoner.

He found the student sitting on the floor of a damp cell. The only illumination came from a small window set high in the concrete wall. Nasadev raised red-rimmed eyes when the door opened, amid a clatter of keys and bolts, and Zakov found himself again amazed at the lad's homeliness, rendered even more pathetic by features wet and swollen from weeping. Somewhat discomfited by his own reaction, he forced a smile.

The young man attempted to rise, but was hindered by the shackles about his ankles.

"Please, remain seated," offered Zakov quickly, and joined him on the earthen floor. "Ludmilla delivered your note. Just what kind of trouble are you in?"

"They are saying that I killed Olga," the youth replied, his face contorted in a grimace. "As if I could! I adored her," he cried. "But things look rather bleak for me, the way they found my knife. Only you can help me, sir. I know your reputation. Please help me, sir! Not for myself, but for my family. This will destroy them..." His tears welled up again, and he sniffled and dabbed vainly at his cheeks with his dirty sleeve.

Zakov looked at him sympathetically. Ludmilla had told him the student was pathetic, and she was right: the ordinary face,

Copyright © Mometrix Media. You have been licensed one copy of this document for personal use only. Any other reproduction or redistribution is strictly prohibited. All rights reserved.
This content is provided for test preparation purposes only and does not imply an endorsement by Mometrix of any particular political, scientific, or religious point of view.

the weak chin, now damp with tears and perspiration, made him seem a most unlikely suitor. He must have been deeply hurt by Olga's rejection. And she, delicate, spirited and appealing, must have found his attentions tedious and annoying, the more so since he had neither fortune nor prospects. Had she derided him, mocked him, gone too far setting trial to his devotion until he killed her?

"What made the police suspect you?" he asked.

"They talked to some of the other students, and my name came up." Nasadev was almost whining. "One of them found a letter I had written, but never sent. He told the others. It made me angry, but I guess it was foolish of me to think that such a girl could ever like someone like me. But she was nice to me, nonetheless. She told me about the play she was writing, and how she hoped to go to Moscow."

Zakov found that his heart went out to the poor, infatuated youth. But pity was not enough to resolve the charges the student faced.

"So the other students mentioned your name," he continued. "That explains why the police spoke to you, but it's not enough for them to bring charges. What else happened?"

Nasadev replied plaintively, fixing Zakov with an imploring look. "We had a quarrel. In the park. The day before they found her. Someone must have heard us and they told the police. They came and searched my room in the dormitory, found my diary, some other letters I had written."

"What was the quarrel about?"

Nasadev hesitated, would not meet Zakov's eye. "I had asked her to marry me. I know it was crazy...how could I support her? But I was afraid she would stop seeing me, that someone else would come along and I would lose her."

"And what did she say?" Zakov knew the answer already, but he had to ask.

"She turned me down. She didn't laugh at me, but she turned me down. And that was in my diary, too." Nasadev was practically inaudible.

Zakov stayed a while longer, asked a few more questions. A deep sadness seemed to fill his heart as he regarded the small cell and its single miserable occupant. Finally, with a shiver, he called for Gorkhi's man to open the door and he left.

22. Zakov thought that Gorkhi, the police chief, was

a. pretentious
b. enthusiastic
c. moderate
d. uncooperative

23. The passage suggests that Nasadev's cell

a. was modern
b. had only a bed in it
c. was spacious
d. was unfurnished

Copyright © Mometrix Media. You have been licensed one copy of this document for personal use only. Any other reproduction or redistribution is strictly prohibited. All rights reserved.
This content is provided for test preparation purposes only and does not imply an endorsement by Mometrix of any particular political, scientific, or religious point of view.

24. Nasadev's frame of mind during the interview can best be described as

a. courageous
b. determined
c. desperate
d. calculating

25. Zakov's attitude toward Nasadev appears to be one of:

a. Disgust
b. Magnanimity
c. Sympathy
d. Malice

26. Nasadev's look, when he replies to Zakov's question (Lines 69–70) implies that he is:

a. Begging for help
b. Looking for words
c. Exhausted
d. Unable to think clearly

27. When the student tells Zakov that Olga's rejection of his proposal was in the diary found by the police (Line 76), his voice:

a. Whines
b. Is hard to hear
c. Contains a signal
d. Breaks

28. By the time Zakov leaves, he

a. has proved Nasadev's guilt
b. is angry at Gorkhi
c. has found a possible motive for Nasadev to have committed the murder
d. has proved Nasadev's innocence

Refer to the following for questions 29–32:

One of the key features of the music scene in the past decade has been the increasing popularity of outsiders, especially those with a career. In previous decades, amateur status was seen as a lower calling or, at best, a step on the way to professional status, but many musical insiders now believe that amateurs actually constitute an elite group within the music scene, with greater chances of eventual success. Professionals, once able to fully devote themselves to the advancement of their musical careers, now find themselves hamstrung by a variety of factors that were not issues even a decade ago, giving the edge to people who do not depend on music for a livelihood. A number of technological, demographic, and economic factors are to blame for this change.

Full-time musicians always had difficulties making ends meet, but these difficulties have been vastly increased by a changing music scene. The increased popularity of electronic music, mega-bands, and other acts that rely heavily on marketing, theatrics, and expensive effects has made it harder than ever for local acts to draw crowds. The decreasing crowds at coffee houses, bars, and other small venues leave the owners without the ability to pay for live music. Amateurs can still play the same

Copyright © Mometrix Media. You have been licensed one copy of this document for personal use only. Any other reproduction or redistribution is strictly prohibited. All rights reserved.
This content is provided for test preparation purposes only and does not imply an endorsement by Mometrix of any particular political, scientific, or religious point of view.

coffee houses as ever, and the lack of a hundred-dollar paycheck at the end of the night is hardly noticed. Professionals, however, have to fight more desperately than ever for those few lucrative gigs.

An even bigger factor has been the rise of digital media in general and digital file sharing in particular. People have been trading copies of music for decades, but in the days of analog tapes there was always a loss. The tape one fan burned for another would be of lesser quality than the original, prompting the recipient to go out and buy the album. Now that music fans can make full-quality copies for little or nothing and distribute them all over the world, it can be very hard for bands to make any money on music sales. Again, this does not make much difference to amateurs, but it robs the professionals of what has traditionally been one of their biggest sources of revenue.

All of this results in a situation so dire for professional musicians that their extra experience often doesn't balance out their lack of economic resources. The amateurs are the only ones who can afford to buy new gear and fix broken equipment, keep their cars in working order to get to shows, and pay to promote their shows. The professionals tend to have to fall back on "day jobs," typically at lower rates and with less opportunity for advancement. Even those professional musicians who are able to supplement their incomes with music lessons, wedding shows, and other traditional jobs are often living at such a low level that they cannot afford to buy the professional equipment they need to keep the higher-paying gigs. A fairly skilled amateur, by contrast, may not have the same level of virtuosity but will be able to fake his way through most of what a professional does at a more competitive rate, which will allow him to play professional shows.

29. The author of this essay is mainly

a. arguing for a return to a climate more favorable to professional musicians.
b. examining the causes of the increasing success of amateur musicians over professionals.
c. revealing the psychological toll the current economy takes on professional musicians.
d. disputing the claim that unsuccessful professional musicians simply don't work hard enough.

30. Which of the following statements about musicians does the essay most directly support?

a. Bars and coffee houses should be willing to pay a fair wage to professional musicians.
b. The most popular professional bands have not been affected by the changes that plague most professional musicians.
c. It is much easier for amateur musicians to book shows than it was a decade ago.
d. Professional musicians have recently lost some of their most important sources of income.

31. In his discussion of professional musicians in the last paragraph, the author

a. indicates that amateurs deserve their new, higher status.
b. shows that in the current climate, professionals may not have the ability to purchase and maintain the tools that they need.
c. points out the decrease in the market for wedding gigs and lessons.
d. predicts a decline in the number of professional musicians.

Copyright © Mometrix Media. You have been licensed one copy of this document for personal use only. Any other reproduction or redistribution is strictly prohibited. All rights reserved.
This content is provided for test preparation purposes only and does not imply an endorsement by Mometrix of any particular political, scientific, or religious point of view.

32. According to the essay, amateur musicians are becoming more successful at both amateur and professional gigs because professionals

a. exclusively performs high-paying gigs and are unwilling to play in clubs.
b. are not able to relate to ordinary people as well as amateurs can.
c. have financial needs that they are not able to meet in the current musical climate.
d. are in an industry that is particularly susceptible to economic changes.

Refer to the following for questions 33–36:

"His pride," said Miss Lucas, "does not offend me so much as pride often does, because there is an excuse for it. One cannot wonder that so very fine a young man, with family, fortune, everything in his favour, should think highly of himself. If I may so express it, he has a right to be proud."

"That is very true," replied Elizabeth, "and I could easily forgive his pride, if he had not mortified mine."

"Pride," observed Mary, who piqued herself upon the solidity of her reflections, "is a very common failing I believe. By all that I have ever read, I am convinced that it is very common indeed, that human nature is particularly prone to it, and that there are very few of us who do not cherish a feeling of self-complacency on the score of some quality or other, real or imaginary. Vanity and pride are different things, though the words are often used synonymously. A person may be proud without being vain. Pride relates more to our opinion of ourselves, vanity to what we would have others think of us."

33. Why doesn't the gentleman's pride offend Miss Lucas?

a. She admires his vanity.
b. It is human nature to be proud.
c. He is poor and homeless.
d. He is handsome and rich.

34. What are Elizabeth's feelings towards the gentleman?

a. She is offended by him.
b. She is proud of him.
c. She wants to get to know him better.
d. She is glad he is rich.

35. Which sentence best states the theme of this passage?

a. Pride and vanity are offensive.
b. Fame and fortune can make a person proud.
c. Every person is proud in one way or another.
d. If you have a fortune, you deserve to be proud.

36. According to the passage, what is the difference between pride and vanity?

a. Pride relates to a person's abilities; vanity relates to a person's looks.
b. Men are proud; women are vain.
c. Pride is what you think of yourself; vanity is what you want others to think of you.
d. Pride is part of human nature; vanity is not.

Copyright © Mometrix Media. You have been licensed one copy of this document for personal use only. Any other reproduction or redistribution is strictly prohibited. All rights reserved.
This content is provided for test preparation purposes only and does not imply an endorsement by Mometrix of any particular political, scientific, or religious point of view.

Answer Key and Explanations #2

Algebra and Functions

1. B: The amount she spends on rent and utilities is equal to 0.38($40,000), or $15,200, which is approximately $15,000.

2. B: To multiply two powers that have the same base, you need to add their exponents. This is represented by the property $x^m \cdot x^n = x^{m+n}$. So, $x^3x^5 = x^{3+5} = x^8$.

3. B: The degree of a polynomial refers to the largest exponent. The degree of $f(x)$ is 1, the degree of $g(x)$ is 2, and the degree of $h(x)$ is 3. The leading coefficient for each function is –2. Functions $f(x)$ and $h(x)$ have exactly one real zero, while $g(x)$ has two real zeros:

$$f(x) = 0 \quad -2x + 1 = 0 \quad -2x = -1 \quad x = \frac{1}{2}$$

$$g(x) = 0 \quad -2x^2 + 1 = 0 \quad -2x^2 = -1 \quad x^2 = \frac{1}{2} \quad x = \sqrt{\frac{1}{2}};\ x = -\sqrt{\frac{1}{2}}$$

$$h(x) = 0 \quad -2x^3 + 1 = 0 \quad -2x^3 = -1 \quad x^3 = \frac{1}{2} \quad x = \sqrt[3]{\frac{1}{2}}$$

The zeros of the function are also the x-intercepts, so none of the functions have an x-intercept at (1,0).

4. C: First, use the properties of logarithms to solve for y:

$2\log_8 y + \log_8 4 = 2$

$\log_8 y^2 + \log_8 4 = 2$	Use the power rule of logarithms to rewrite the first logarithm.
$\log_8 4y^2 = 2$	Use the product rule of logarithms to combine the two logarithms.
$4y^2 = 8^2$	Rewrite the equation using the logarithm rule that $\log_a b = c$ is equivalent to $a^c = b$.
$4y^2 = 64$	Simplify the square on the right side.
$y^2 = 16$	Divide both sides by 4.
$y = 4$	Take the positive square root of both sides.

Finally, substitute 4 for y in the expression $\log_y 64$ and simplify: $\log_4 64 = 3$ since $4^3 = 64$.

Copyright © Mometrix Media. You have been licensed one copy of this document for personal use only. Any other reproduction or redistribution is strictly prohibited. All rights reserved.
This content is provided for test preparation purposes only and does not imply an endorsement by Mometrix of any particular political, scientific, or religious point of view.

5. B: First, state the exclusions of the domain. This can be done by factoring the denominator, then finding where the denominator is equal to 0.

$$x^3 + 2x^2 - x - 2 \neq 0$$
$$x^2(x+2) - (x+2) \neq 0$$
$$(x+2)(x^2 - 1) \neq 0$$
$$(x+2)(x-1)(x+1) \neq 0$$
$$x + 2 \neq 0 \quad x - 1 \neq 0 \quad x + 1 \neq 0$$
$$x \neq -2 \qquad x \neq 1 \qquad x \neq -1$$

To determine whether there are vertical asymptotes or holes at these values of x, simplify the expression:

$$\frac{x^2 - x - 6}{x^3 + 2x^2 - x - 2} = \frac{(x-3)(x+2)}{(x+2)(x-1)(x+1)} = \frac{x-3}{(x-1)(x+1)}$$

There are vertical asymptotes at $x = 1$ and at $x = -1$ and a hole at $x = -2$. Statement I is false.

To find the x-intercept of $f(x)$, solve $f(x) = 0$. $f(x) = 0$ when the numerator is equal to zero.

$$x^2 - x - 6 = 0$$

$$(x+2)(x-3) = 0$$

$$x = -2,3$$

The numerator equals zero when $x = -2$ and $x = 3$; however, -2 is excluded from the domain of $f(x)$, so the x-intercept is 3. To find the y-intercept of $f(x)$, find $f(0)$.

$$\frac{(0)^2 - (0) - 6}{(0)^3 + 2(0)^2 - (0) - 2} = \frac{-6}{-2} = 3$$

The y-intercept is 3. Statement II is true.

6. B: When $y = x^3$, $x = \sqrt[3]{y}$. Similarly, when $y = e^x$, $x = \ln y$ for $y > 0$. On the other hand, when $y = x + a$, $x = y - a$; when $y = \frac{1}{x}$, $x = \frac{1}{y}$ for $x, y \neq 0$; and when $y = \sin x$, $x = \sin^{-1} y$.

7. D: Lines with the same slope will be parallel. The equation $y = 8x - 4$ is in slope-intercept form, $y = mx + b$, where m represents the slope and b represents the y-intercept. The slope in the equation $y = 8x - 4$ is 8. The only equation listed with a slope of 8 is $y = 8x - 1$. The lines $y = 8x - 1$ and $y = 8x - 4$ will be parallel when graphed.

8. B: By definition, when $f(x)$ and $g(x)$ are inverse functions, $f(g(x)) = g(f(x)) = x$. So, $f\left(g\left(\frac{1}{8}\right)\right) = x = \frac{1}{8}$.

9. C: We are looking for a graph with two key pieces of information: rainfall every day except Friday, and an amount of rain on Monday double the rain on Thursday. Answer choice A is incorrect because it shows no rain on Sunday. Answer choice B is incorrect because it shows rain on Friday. Answer choice D is incorrect because the amount of rain on Monday is equal to the amount of rain

Copyright © Mometrix Media. You have been licensed one copy of this document for personal use only. Any other reproduction or redistribution is strictly prohibited. All rights reserved.
This content is provided for test preparation purposes only and does not imply an endorsement by Mometrix of any particular political, scientific, or religious point of view.

on Thursday. Only answer choice C is correct because it shows no rain on Friday, 6 inches on Monday, and 3 inches on Thursday.

10. B: To simplify, multiply the value outside of the parentheses by each value inside of the parentheses.

$$9x(3x^2 + 2x - 9) = 9x(3x^2) + 9x(2x) + 9x(-9)$$
$$= 27x^3 + 18x^2 - 81x$$

11. D: To factor this equation, we need to think of things using the reverse of the FOIL method. Start by setting up two empty parentheses.

$$3x^2 + 4x - 15 = (\quad)(\quad)$$

First, determine what goes in the first part of each set of parentheses. The only way to multiply with integers and get $3x^2$ is $3x \cdot x$. Put these terms in each of the first parts of the sets of parentheses.

$$3x^2 + 4x - 15 = (3x \quad)(x \quad)$$

Next, determine what numbers multiply to –15 and will make the middle term of the multiplied expression equal to $4x$.

$$3x^2 + 4x - 15 = (3x - 5)(x + 3)$$

Notice that it is important to make sure –5 goes in the left set of parentheses and +3 goes in the right set of parentheses. If they were switched, using the FOIL method on the two binomials would result in $(3x + 3)(x - 5) = 3x^2 - 12x - 15$, which is not the desired expression. Therefore, the correct factorization of the expression is $(3x - 5)(x + 3)$.

12. B: Lines will be perpendicular if one slope is the negative reciprocal of the other slope. The equation $y = 0.5x + 7.4$ is in slope-intercept form, $y = mx + b$, where m represents the slope and b represents the y-intercept. The slope of $y = 0.5x + 7.4$ is 0.5, or $\frac{1}{2}$. The negative reciprocal of $\frac{1}{2}$ is -2. The only equation listed with a slope of –2 is $y = -2x + 5.5$. This line will be perpendicular to the line of the original equation $y = 0.5x + 7.4$.

13. B: This pie chart shows the percentage of the total recyclable material that each material represents. The larger percentages have larger slices of the circle. Also, the percentage for each material is shown next to each slice. In this chart, paper is the most recycled material because it has the largest slice. This is 40% of the total. The next most common is glass at 25% of the total. All of the other materials stand for smaller portions of the total.

14. C: The chart indicates that 40% of the total recycled material is paper. Since 50,000 tons of material are recycled every month, the total amount of paper will be 40% of 50,000 tons.

$$40\% \times 50{,}000 = 0.4 \times 50{,}000 = 20{,}000$$

Therefore, about 20,000 tons of paper are recycled each month.

15. B: The vertex form of a quadratic equation is $y = a(x - h)^2 + k$, where $x = h$ is the parabola's axis of symmetry and (h, k) is the parabola's vertex. The vertex of the graph is (1,2), so the equation

Copyright © Mometrix Media. You have been licensed one copy of this document for personal use only. Any other reproduction or redistribution is strictly prohibited. All rights reserved.
This content is provided for test preparation purposes only and does not imply an endorsement by Mometrix of any particular political, scientific, or religious point of view.

can be written as $y = a(x-1)^2 + 2$. The parabola passes through the point (0,1), so $1 = a(0-1)^2 + 2$. Solve for a:

$$1 = a(0-1)^2 + 2$$
$$1 = a + 2$$
$$-1 = a$$

So, the vertex form of the parabola is $y = -(x-1)^2 + 2$. Write the equation in standard form: $y = ax^2 + bx + c$.

$$\begin{aligned} y &= -(x-1)^2 + 2 \\ y &= -(x-1)(x-1) + 2 \\ y &= -(x^2 - 2x + 1) + 2 \\ y &= -x^2 + 2x - 1 + 2 \\ y &= -x^2 + 2x + 1 \end{aligned}$$

16. D: Recall that the general form of a quadratic expression is $ax^2 + bx + c$. A great way to factor quadratic expression like this, where $a = 1$ and all the answer choices are integer factors, would be to consider the factors of the last term, c. Specifically, any two factors of c that would add to b. Essentially: $f_1 \times f_2 = -12$ and $f_1 + f_2 = 1$. We can check the factors of -12.

f_1	f_2	$f_1 + f_2$
12	-1	11
6	-2	4
4	-3	1
3	-4	-1
2	-6	-4
1	-12	-11

From this the only option that works is 4 and -3, which means the expression factors as $(x + 4)(x - 3)$.

17. A: One way to determine which is the graph of $f(x) = -2|-x+4| - 1$ is to find $f(x)$ for a few values of x. For example, when $x = 0$, $f(0) = -2|0+4| - 1 = -9$. Graphs A and B pass through $(0, -9)$, but graphs C and D do not. When $x = 4$, $f(4) = -2|-4+4| - 1 = -1$. Graph A passes through $(4, -1)$, but graph B does not. Graph A is the correct graph.

18. A: The new line needs to have a slope that is the negative reciprocal of the original slope, -5. The reciprocal of -5 is $-\frac{1}{5}$. When this is negated, it becomes $\frac{1}{5}$. This means that the negative reciprocal of -5 is $\frac{1}{5}$. The slope of the new line needs to be $\frac{1}{5}$.

19. D: Remember that when you multiply like bases, you add the exponents, and when you divide like bases, you subtract the exponents.

$$(xy)^{7y} - (xy)^y = (xy)^y[(xy)^{7y-y} - 1] = (xy)^y[(xy)^{6y} - 1]$$

20. A: This equation is a linear relationship that has a slope of 3.60 and passes through the origin. The table shows that for each hour of rental, the cost increases by \$3.60. This matches with the slope of the equation. Of course, if the bicycle is not rented at all (0 hours), there will be no charge

Copyright © Mometrix Media. You have been licensed one copy of this document for personal use only. Any other reproduction or redistribution is strictly prohibited. All rights reserved.
This content is provided for test preparation purposes only and does not imply an endorsement by Mometrix of any particular political, scientific, or religious point of view.

(\$0). If plotted on the Cartesian plane, the line would have a y-intercept of 0. Choice A is the only one that meets these requirements.

21. C: To solve, use the percentage increase formula.

$$\text{Percentage Increase} = \frac{\text{new} - \text{initial}}{\text{initial}} \times 100$$

In this case, the initial value is \$80, and the new value is \$120.

$$\text{Percentage Increase} = \frac{120 - 80}{80} \times 100 = \frac{40}{80} \times 100 = 50\%$$

Therefore, the total amount in her account increased by 50%.

22. C: The equation can be translated:

$$\text{height} = (\text{original height when planted}) + 0.75(\text{number of years since planted})$$

In other words, it grows 0.75 feet every year. The question asks for the number of inches, so we multiply 0.75 by 12 to obtain 9 inches per year.

23. B: Substitute the given values and solve. Simplify the operations inside parentheses first.

$$\begin{aligned}\frac{a(b-c)}{b(a+b+c)} &= \frac{4(3-1)}{3(4+3+1)} \\ &= \frac{4(2)}{3(8)} \\ &= \frac{8}{24} \\ &= \frac{1}{3}\end{aligned}$$

24. A: By definition, when $f(x)$ and $g(x)$ are inverse functions, $f(g(x)) = g(f(x)) = x$. So, $f(g(4)) = b = -6$.

25. A: The vertex form of a quadratic equation is $y = a(x-h)^2 + k$, where $x = h$ is the parabola's axis of symmetry and (h, k) is the parabola's vertex. The vertex of the graph is $(-2, -1)$, so the equation can be written as $y = a(x+2)^2 - 1$. The parabola passes through point $(0,3)$, so $3 = a(0+2)^2 - 1$. Solve for a.

$$\begin{aligned}3 &= a(0+2)^2 - 1 \\ 3 &= a(2)^2 - 1 \\ 3 &= 4a - 1 \\ 4 &= 4a \\ 1 &= a\end{aligned}$$

Copyright © Mometrix Media. You have been licensed one copy of this document for personal use only. Any other reproduction or redistribution is strictly prohibited. All rights reserved.
This content is provided for test preparation purposes only and does not imply an endorsement by Mometrix of any particular political, scientific, or religious point of view.

So, the vertex form of the parabola is $y = (x + 2)^2 - 1$. Write the equation in standard form, $y = ax^2 + bx + c$.

$$y = (x + 2)^2 - 1$$
$$y = (x + 2)(x + 2) - 1$$
$$y = (x^2 + 4x + 4) - 1$$
$$y = x^2 + 4x + 3$$

26. D: Set the given expression equal to y, then solve for y. When $y = \frac{1}{\sqrt{x}}$, $x = \frac{1}{y^2}$ or y^{-2}. Similarly, when $y = \cos x$, $x = \cos^{-1} y$. On the other hand, when $y = 2x$, $x = \frac{y}{2}$; when $y = e^x$, $x = \ln y$ for $y > 0$; and when $y = \frac{1}{x^2}$, $x = \pm\sqrt{\frac{1}{y}}$.

27. C: First, set the equation equal to zero, then factor it.

$$2x^3y^2 - 12x^2y = 4x^3y - 6x^2y^2$$
$$2x^3y^2 - 4x^3y + 6x^2y^2 - 12x^2y = 0$$
$$2x^2y(xy - 2x + 3y - 6) = 0$$
$$2x^2y[x(y - 2) + 3(y - 2)] = 0$$
$$2x^2y[(y - 2)(x + 3)] = 0$$

Use the zero product property to find the solutions.

$2x^2y = 0$	$y - 2 = 0$	$x + 3 = 0$
$x = 0$	$y = 2$	$x = -3$
$y = 0$		

So, the solutions are $x = \{0, -3\}$ and $y = \{0,2\}$.

28. A: A system of linear equations can be solved by using matrices or by using the graphing, substitution, or elimination method. The elimination method is shown here.

$$\begin{cases} 3x + y = 1 \\ -4x + 3y = 29 \end{cases}$$

To eliminate y by the elimination method, multiply the top equation by –3 so that the coefficients of the y-terms will be additive inverses.

$$-3(3x + y) = (1)(-3)$$
$$-9x - 3y = -3$$

Then, add the two equations and solve for x.

$$\begin{array}{rrcr} & -9x \;\; -3y & = & -3 \\ + & -4x \;\; +3y & = & 29 \\ \hline & -13x & = & 26 \\ & x & = & -2 \end{array}$$

Copyright © Mometrix Media. You have been licensed one copy of this document for personal use only. Any other reproduction or redistribution is strictly prohibited. All rights reserved.
This content is provided for test preparation purposes only and does not imply an endorsement by Mometrix of any particular political, scientific, or religious point of view.

Substitute -2 for x in either of the given equations and solve for y. Here $x = -2$ is substituted into the top equation.

$$3(-2) + y = 1$$
$$-6 + y = 1$$
$$y = 7$$

The solution to the system of equations is $(-2, 7)$.

29. A: The inequality that represents this scenario is $n + 3p \leq 30$, where n is the number of notebooks and p is the number of boxes of pencils. To graph this inequality, find the intercepts of this linear inequality by making the inequality an equality and setting n and p equal to 0 and solving for the other variable.

$$(0) + 3p = 30 \qquad n + 3(0) = 30$$

$$3p = 30 \qquad n = 30$$

$$p = 10$$

The intercepts are $n = 30$ and $p = 10$. The solid line through the two intercepts represents the maximum number of each item that may be purchased. Since this inequality represents the maximum he can purchase, all first-quadrant points below the line are also possible amounts within Ben's budget.

30. B: A simple test of each option at $x = -5$, $x = 0$, and $x = 2$ demonstrates that only choice B works:

x	**Observed**	y_a	y_b	y_c	y_d
−5	$\sim \frac{1}{4}$	−2	$\frac{1}{6}$	$-\frac{7}{6}$	$\frac{1}{9}$
0	$\sim -\frac{1}{4}$	$\frac{1}{2}$	$-\frac{1}{4}$	$-\frac{3}{4}$	$-\frac{1}{6}$
2	$\sim \frac{1}{4}$	$\frac{1}{3}$	$\frac{1}{6}$	$\frac{7}{6}$	$\frac{1}{9}$

31. C: We can look at each equation separately.

Choice A, $y = \frac{1}{x-1} + 1$, has a zero denominator at $x = 1$, so we can eliminate it immediately.

Choice B, $y = -\sqrt{x+2} - 1$, does include $x = 1$ in its domain: $y = -\sqrt{(1) + 2} - 1 \approx -2.73$. But we can see that all y-values will be a negative number (or zero) minus 1, which means all y-values will be negative, so 2 is outside the range.

Copyright © Mometrix Media. You have been licensed one copy of this document for personal use only. Any other reproduction or redistribution is strictly prohibited. All rights reserved.
This content is provided for test preparation purposes only and does not imply an endorsement by Mometrix of any particular political, scientific, or religious point of view.

Choice C, $y = |x + 2| - 3$, also includes $x = 1$ in its domain: $y = |(1) + 2| - 3 = 0$. And we can see that $y = |(3) + 2| - 3 = 2$, so 2 is in the range. So choice C meets both criteria.

Choice D also includes $x = 1$ in its domain: $y = -(1) - 3 = -4$. But by graphing this piecewise linear function, we can see that its range is $y < -1$, which excludes 2.

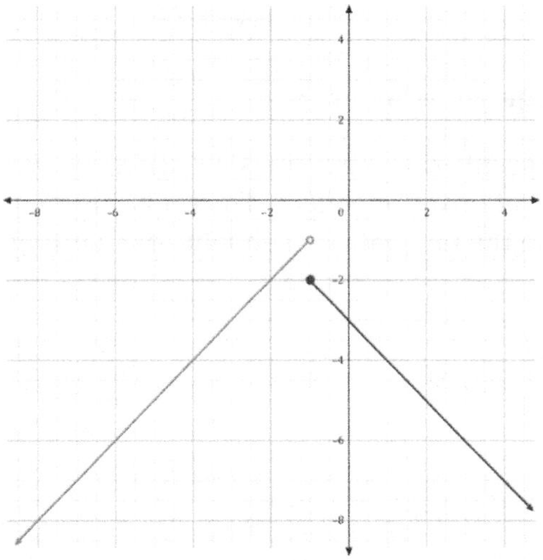

32. D: There are 90 two-digit numbers (all integers from, and including, 10 to 99). Of those, there are 13 multiples of 7: 14, 21, 28, 35, 42, 49, 56, 63, 70, 77, 84, 91, 98. Therefore, the probability of choosing a number that is a multiple of 7 is $\frac{13}{90}$.

33. D: To add the two fractions, first rewrite them with the least common denominator, which is in this case y^3. This is already the denominator in $\frac{x}{y^3}$, and we can rewrite $\frac{x^2}{y^2}$ as $\frac{x^2 \times y}{y^2 \times y} = \frac{x^2 y}{y^3}$. Thus, $\frac{x^2}{y^2} + \frac{x}{y^3} = \frac{x^2 y}{y^3} + \frac{x}{y^3} = \frac{x^2 y + x}{y^3}$.

Reading Comprehension

1. C: The first paragraph gives the information to correctly answer this question.

2. B: This passage in arranged by problem and solution. The author states a problem that the cotton farmers were having: "Eventually, cotton will no longer grow on this land." The author then presents a solution: "Carver told farmers to rotate their crops: plant cotton one year, then the next year plant peanuts and other soil-restoring plants, like peas and sweet potatoes. It worked! The peanut plants grew and produced lots of peanuts."

3. D: The second paragraph discusses the problem the cotton farmers were facing. The cotton crops had depleted the nutrients from the soil.

4. A: The last paragraph answers this question. "Carver told farmers to rotate their crops: plant cotton one year, then the next year plant peanuts and other soil-restoring plants, like peas and sweet potatoes. It worked! The peanut plants grew and produced lots of peanuts."

5. B: Answer choice B best summarizes what this passage is mainly about. Choices A, and C are not even discussed in this passage. Paragraph 5 does discuss choice D, but it is not the main focus of the passage.

Copyright © Mometrix Media. You have been licensed one copy of this document for personal use only. Any other reproduction or redistribution is strictly prohibited. All rights reserved.
This content is provided for test preparation purposes only and does not imply an endorsement by Mometrix of any particular political, scientific, or religious point of view.

6. A: Answer choice A best defines "crop rotation," as the passage gives the definition in the last paragraph.

7. B: As set out in section 1.0, Purpose and Scope, the document describes a procedure for identifying hazards associated with one or more jobs (in this case Tank Operations) or encountered by visitors and for instituting controls to mitigate (or minimize) the dangers that they present.

8. D: The methods to be used to mitigate hazards are given in section 4.1 of the text, which indicates that they are specified in descending order of use. Protective clothing is an example of the last method listed, personal protective equipment, so this is the last strategy to be tried to protect the workers from the hazardous chemical.

9. D: Section 4.2 indicates that safety procedures ("controls") that fall within the scope of normal training for workers do not need to be discussed in the work instructions for operations that are performed frequently.

10. A: The document is intended as a guide for those writing work instructions for jobs to be performed at the site. These work instructions are prepared by management. Workers and laborers would read those documents as part of their training.

11. D: Section 4.5 indicates that safety procedures that are not within the qualification and training of the workers for hazards should have detailed instructions for how the workers are to mitigate the hazard.

12. C: The passage makes no mention of metals or horses. Although we may infer that they hunted the buffalo because they were plentiful, that is not stated in the passage.

13. B: Travail means work, or effort, and shows that the crows made it more difficult for the people to kill buffalo during the hunt.

14. B: The story tells us that after the great white crow turned black, all the other crows were black as well. Thus, he is a symbol for all these birds.

15. D: Lines 29–30 tells us that the tribe planned to frighten the chief of the crows to prevent the crows from warning the buffalo about the hunts. The passage does not suggest that they hated all birds or that they planned to eat this one.

16. C: Long Arrow acted like the buffalo in the herd so that the chief of the crows would approach, making it possible to capture him. Although we may infer that he had to fool the buffalo in the herd as well, this is secondary to his need to fool the birds.

17. B: These details help us to see how the people lived. Although they hunted with the stone-tipped spears, the rawhide bag was not a part of the hunt.

18. A: As he lands, he asks "have you not heard my warning?" (Lines 68–69).

19. A: The suggestions included several for killing or mutilating the bird, which does not suggest a calm resolve. And there is no suggestion that they were either celebrating or hungry at this time.

20. D: There is no characterization of Long Arrow in the passage, and we know nothing about him or why he was chosen.

Copyright © Mometrix Media. You have been licensed one copy of this document for personal use only. Any other reproduction or redistribution is strictly prohibited. All rights reserved.
This content is provided for test preparation purposes only and does not imply an endorsement by Mometrix of any particular political, scientific, or religious point of view.

21. B: The birds in the story are able to observe the actions of hunters, to interpret them as potentially harmful for their buffalo friends, and to act for the protection of the buffalo. They do not appear to do this for their own benefit, nor do they seem to act specifically to harm the tribe, but rather to help the buffalo.

22. A: Gorkhi is described in line 5 as a bit bombastic. A bombastic individual is one who is overbearing and pretentious.

23. C: Lines 10–11 tells us that Zakov found the student on the floor.

24. D: Nasadev is described as whining, with his eyes red from crying, and concerned that his troubles will affect his family.

25. C: Zakov tries to understand how the student must have felt about Olga's rejection. His "heart went out" to the student, and later, when he leaves, he feels a deep sadness for him.

26. A: The student regards Zakov with an imploring look. To implore is to beg.

27. B: The passage describes his voice as "practically inaudible," which means "unable to be heard," but does not describe it as a whisper.

28. C: Olga's rejection of the student provides a reasonable motive, but there is nothing in the passage to suggest that Zakov had proof of Nasadev's guilt or innocence.

29. B: In the first paragraph of the essay, the author characterizes amateurs as "an elite group within the music scene" and states that there are several "technological, demographic, and economic factors" that account for them doing better than professionals. The tone of the essay is documentary—the author doesn't make any judgments about whether this is a good development or a bad one. He simply states that amateurs are more successful relative to professionals than they have been before and goes on to examine the reasons for this. Therefore, choice B is correct.

30. D: The key is the phrase "directly support." The essay needs to come right out and say the correct answer, not imply that it is true. Paragraph 3 says that digital file sharing "robs the professionals of what has traditionally been one of their biggest sources of revenue." Paragraph 2 provides less direct evidence, saying that many clubs that were once able to pay professionals now can't. Professionals have lost most of their income from both small clubs and recordings. Therefore, choice D is correct.

31. B: The second sentence in the final paragraph is a giveaway. If the "amateurs are the only ones who can afford to buy new gear and fix broken equipment, keep their cars in working order to get to shows, and pay to promote their shows," then the professionals must not be able to do any of those things, as stated in choice B.

32. C: The essay as a whole discusses how the current musical scene negatively affects professional musicians while leaving amateurs unharmed. The second paragraph, for example, discusses how professionals are no longer able to make a living playing small venues and must "fight more desperately than ever for those few lucrative gigs." The final paragraph states that, because of the effect on their finances, professionals are unable to maintain the gear and transportation they need to "keep the higher-paying gigs." It goes on to say that "a fairly skilled amateur . . . will be able to fake his way through most of what a professional does . . . to play professional shows." Therefore, professionals are falling behind amateurs at small venues (which professionals can't afford to play

Copyright © Mometrix Media. You have been licensed one copy of this document for personal use only. Any other reproduction or redistribution is strictly prohibited. All rights reserved.
This content is provided for test preparation purposes only and does not imply an endorsement by Mometrix of any particular political, scientific, or religious point of view.

because of the lack of pay) and at professional gigs (where professionals can't play because they can't afford professional gear). Therefore, choice C is correct.

33. D: In the first paragraph, Miss Lucas states that "so very fine a young man, with family, fortune, every thing in his favour, should think highly of himself. If I may so express it, he has a *right* to be proud." Basically, she feels he deserves to be proud because he is physically attractive, comes from a good family, has money, and is successful.

34. A: This question is asking you to make an inference about Elizabeth's feeling towards the gentleman. In paragraph 2, Elizabeth is "mortified" by the gentleman's actions towards her. From this statement, you can make the inference that she was offended by his actions.

35. C: Theme is a message or lesson conveyed by a written text. The message is usually about life, society or human nature. This particular excerpt is exploring pride as it relates to human nature. Mary's observations on pride are the best summary of the theme of this passage. "By all that I have ever read, I am convinced that it is very common indeed, that human nature is particularly prone to it." The best answer is choice C.

36. C: Paragraph 3 gives the answer to this question. According to Mary, pride is an opinion of yourself, and vanity is what we want others to think of us.

Copyright © Mometrix Media. You have been licensed one copy of this document for personal use only. Any other reproduction or redistribution is strictly prohibited. All rights reserved.
This content is provided for test preparation purposes only and does not imply an endorsement by Mometrix of any particular political, scientific, or religious point of view.

Additional Bonus Material

Due to our efforts to try to keep this book to a manageable length, we've created a link that will give you access to all of your additional bonus material:

mometrix.com/bonus948/ibewapt

Copyright © Mometrix Media. You have been licensed one copy of this document for personal use only. Any other reproduction or redistribution is strictly prohibited. All rights reserved.
This content is provided for test preparation purposes only and does not imply an endorsement by Mometrix of any particular political, scientific, or religious point of view.

Made in the USA
Monee, IL
16 November 2024

70309744R00111